WINDS OF CHANGE
EAST ASIA'S SUSTAINABLE ENERGY FUTURE

WINDS OF CHANGE
EAST ASIA'S SUSTAINABLE ENERGY FUTURE

Xiaodong Wang

Noureddine Berrah

Subodh Mathur

Ferdinand Vinuya

 THE WORLD BANK

ISBN: 978-0-8213-8486-2
eISBN: 978-0-8213-8502
DOI: 10.1596/978-0-8213-8486-2

Library of Congress Cataloging-in-Publication data has been requested.

Cover photographs:
Left: David Llorito, World Bank, Manila office. Man installing energy-efficient light bulb, the Philippines.
Right: Anissa Tria, World Bank. Young girl standing on windfarm, Northwind Bangui Bay Project, Ilocos Norte, the Philippines.

Cover/book design and typesetting: BMW&W Publishing Services, Baltimore, MD.

Contents

Boxes

Foreword

The World Development Report 2010: Development and Climate Change (WDR) concludes that a "climate-smart" world is within reach, if we act now, act together, and act differently from the past. This report demonstrates that a "climate-smart" energy strategy is possible for countries in the East Asia region, with support from the international community.

In the past three decades, the East Asia region has experienced the fastest economic growth in the world, accompanied by rapid urbanization. As a consequence, energy consumption has more than tripled and is expected to further double over the next two decades. This remarkable growth and rapid urbanization have led to twin energy challenges in the region: improving environmental sustainability and enhancing energy security. The region has many of the world's most polluted cities, resulting from fossil fuel combustion. The region also contains some of the largest greenhouse gas emitters in the world, although their per capita and historical emissions are much below the levels of industrialized countries. Concerns with energy security have grown because of increased risks of price volatility and possible disruptions in supplies for oil and gas.

This report focuses on East Asia's sustainable energy development in its middle-income countries: China, Indonesia, Malaysia, the Philippines, Thailand, and Vietnam. A separate report is being prepared on the expansion of energy access and improvement of fuel security in low-income countries in the East Asia and Pacific region: Cambodia, the Lao People's Democratic Republic, Mongolia, Papua New Guinea, the Pacific Islands, and Timor-Leste.

This report suggests the strategic direction for the region to get onto a sustainable energy path over the next two decades, and presents policy tools and financing mechanisms to get there. The main conclusion is that such a path of maintaining economic growth, mitigating climate change, and improving energy security is within reach of the region's countries.

However, what is required is a paradigm shift to a new low-carbon development model. Countries need to act now to transform the energy sector toward much higher energy efficiency and more widespread deployment of low-carbon technologies. At the same time, the speed and scale of urbanization in the region presents an unrivaled opportunity to build low-carbon cities by integrating sustainable energy technologies, compact urban planning, water management, and public transport. While many countries in the region are already taking steps in this direction, accelerating the speed and scaling up the efforts are needed to reach a sustainable energy path.

To move the region to a sustainable energy path, the commitment of the respective governments and communities is essential. The governments will need energy-pricing reforms that no longer encourage the use of fossil fuels, and put in place regulations and incentives that improve energy efficiency and support low-carbon technologies. The governments also will need to ramp up research and development for new technologies to leapfrog to the clean energy revolution.

The countries cannot move to a sustainable energy path alone. They will need the support of the international community. Substantial concessional financing is essential to motivate energy efficiency and low-carbon technology investments. Transfer of low-carbon technologies and institutional strengthening also will be needed.

The World Bank Group is committed to support the governments in the region to make this shift toward a sustainable energy path. The World Bank Group will better integrate our various financial resources to increase our support to energy efficiency, renewable energy, and new technologies; and to coordinate with our development partners to meet the region's needs more effectively. The World Bank Group also will take the initiative to provide the global knowledge and institutional support that the countries need to make this happen.

The World Bank Group looks forward to working with our client countries in the region to bring about the changes that will make a difference locally and globally.

James W. Adams
Vice President
East Asia and Pacific Region
The World Bank Group

Acknowledgments

This flagship report has been prepared by a core team led by Xiaodong Wang and comprising Ranjit Lamech (former Task Team Leader), Joel Maweni (former Task Team Leader), Noureddine Berrah, Subodh Mathur, Ferdinand Vinuya, and Shawna Fei Li. Economic Consulting Associates (ECA) developed the Policy Assessment Model (PAM) and provided consulting services. John Rogers, Georges Darido, and the consulting firms of Almec, MVA, and Segment Y. conducted the urban transport study. Alberto Ang Co, Gouthami Padam, and Ferdinand Vinuya conducted the household study. Ximing Peng and Ming Hu evaluated the modelling work. Alicia Hetzner edited the report.

This work was conducted under the guidance of John Roome, Vijay Jagannathan, and Dejan Ostojic. James Adams, Katherine Sierra, and Jamal Saghir provided valuable comments and guidance.

The team wishes to acknowledge those inside and outside the World Bank Group who contributed input and comments: Ahmad Ahsan, Robert Bacon, Mara Baranson, Fatih Birol, Jan Bojo, Alan Coulthart, Arish Dastur, Philippe Dongier, Gailius Draugelis, Charles Feinstein, Stephen Howes, Ede Jorge Ijjasz-Vasquez, Migara Jayawardena, Martin Joerss, Anthony Jude, Masami Kojima, Feng Liu, Kseniya Lvovsky, Alan Miller, Lucio Monari, Jostein Nygard, Haifeng Qian, Ashok Sarkar, Jas Singh, Rajendra Singh, Richard Spencer, Robert Taylor, Jonathan Walters, Leiping Wang, and Yabei Zhang.

The team benefited greatly from a wide range of consultations in China, Indonesia, Thailand, and Vietnam. The team wishes to thank the participants in these workshops, who included academicians, policy researchers, government officials, members of nongovernmental and civil society organizations, the private sector, and donors.

Finally, the team wishes to acknowledge the generous support from the Governments of Australia and Japan, the Asia Sustainable and Alternative Energy Program (ASTAE), and the Energy Sector Management Assistance Program (ESMAP).

Acronyms and Abbreviations

AAA	Analytical and advisory activities
ADB	Asian Development Bank
ALT	Alternative scenario
APEC	Asia-Pacific Economic Cooperation
APERC	Asia Pacific Energy Research Centre
ASTAE	Asia Sustainable and Alternative Energy Program
AUV	Asian utility vehicle
BAU	Business-as-usual
BBL	Billion barrels
bcm	Billion cubic meter
bil	Billion
BMA	Bangkok Metropolitan Administration
BOT	Build-operate-transfer
CAFE	Corporate automobile fuel economy (standards)
CCS	Carbon capture and storage
CDM	Clean Development Mechanism
CEEP	Chiller Energy Efficiency Project
CFL	Compact fluorescent lamp
CNG	Compressed natural gas
CO	Carbon monoxide
CO_2e	Carbon dioxide equivalent
CRESP	China Renewable Energy Scale-up Program
CTF	Clean Technology Fund
DSM	Demand side management
EAP	East Asia and Pacific (Region, World Bank)
EASEF	East Asia Sustainable Energy Forum
ECA	Economic Consulting Associates
ECCJ	Energy Conservation Centre of Japan
EE	Energy efficiency
EESI	Environmental and Energy Study Institute
EGAT	Electricity Generating Authority of Thailand
EIA	Energy Information Authority
EMC	Energy Management Co.
EMCA	Energy Management Company Association
ESCO	Energy service company
ESMAP	Energy Sector Management Assistance Program

ETAAC	Economic and Technology Advancement Advisory Committee
ETTV	Envelope thermal transfer value
EU	European Union
FDI	Foreign direct investment
GDP	Gross domestic product
GEF	Global Environment Facility
GHG	Greenhouse gas
GJ	Gigajoule
Gt	Gigaton
GW	Gigawatt
GWh	Gigawatt hour
HCMC	Ho Chi Minh City
HSBC	Hong Kong and Shanghai Banking Corporation
IBRD	International Bank for Reconstruction and Development
IDA	International Development Agency
IEA	International Energy Agency
IFC	International Finance Corporation
IFI	International financial institution
IGCC	Integrated gasification combined cycle
IIASA	International Institute for Applied System Analysis
IPCC	Intergovernmental Panel on Climate Change
IS	Inverse Simpson
kWh	Kilowatt hour
LNG	Liquefied natural gas
LPG	Liquefied petroleum gas
MDG	Millennium Development Goal
METI	Ministry of Economy, Trade and Industry
MIC	Middle-income country
M$GDP	Gross domestic product in US$ million
MIT	Massachusetts Institute of Technology
MLF	Multilateral Fund
mmBtu	Million British thermal units
MOF	Ministry of Finance
MP	Master plan
MPG	Miles per gallon
MPV	Multiple person vehicle
Mt	Million tons
$MtCO_2$	Million tons of CO_2
Mtoe	Million tons of oil equivalents
MUV	Multi-use vehicle
MVA	Systra MVA Consulting
MW	Megawatt
NOC	National Oil Company
NO_x	Nitrous oxides
NPC	National Power Company
NPV	Net present value
NT2	Nam Theun 2 Hydroelectric Project
NYSERDA	New York Energy Research and Development Authority

O&M	Operations and maintenance
OECD	Organisation for Economic Co-operation and Development
PAM	Policy Assessment Model
PC	Passenger car
PDP	Power Development Plan
PIC	Pacific Island Country
PM10	Particulate matter, less than 10 microns
PM2.5	Particulate matter, less than 2.5 microns
PNOC-EDC	Philippine National Oil Company Energy Development Corporation
PPA	Power purchase agreement
PPP	Purchasing power parity
PV	Photovoltaic
R&D	Research and development
REDP	Renewable Energy Development Project
REF	Reference scenario
RD&D	Research, development, and demonstration
Rp	Rupiah
RPSs	Renewable portfolio standards
SME	Small and medium-sized enterprise
SED	Sustainable energy development scenario
SO_2	Sulfur dioxide
2W	2-wheeled vehicles
TA	Technical assistance
tCO_2	Tons of CO_2
TDM	Transportation demand management
toe	Ton of oil equivalent
TPES	Total primary energy supply
TSP	Total suspended particulates
TT	Technology transfer
UCT	Unconditional cash transfer
UK	United Kingdom
UNFCCC	United Nations Framework Convention on Climate Change
UNFPA	United Nations Population Fund
USC	Ultrasupercritical
VKT	Vehicle kilometers traveled
VOC	Volatile organic compound
WBG	World Bank Group
WBCSD	World Business Council for Sustainable Development
WDR	*World Development Report*
WEF	World Economic Forum
WEO	World Energy Outlook
WHO	World Health Organization
WRI	World Resources Institute
$bil	Billion US dollars
$M	Million US dollars

Executive Summary

The key messages for this report are:

- **It is within the reach of East Asia's governments to maintain economic growth, mitigate climate change, and improve energy security.** The underlying study for this report found that large-scale deployment of energy efficiency and low-carbon technologies can simultaneously stabilize East Asia's CO_2 emissions by 2025 and significantly improve the local environment and enhance energy security, without compromising economic growth.

- **To achieve these goals requires that governments take immediate action to transform their energy sectors toward much higher energy efficiency and more widespread use of low-carbon technologies.** While many East Asian countries are taking steps in these directions, accelerating the speed and scaling up the efforts are needed to get onto a sustainable energy path. The window of opportunity is closing fast, because delaying action would lock the region into a long-lasting high-carbon infrastructure.

- **This shift to a clean energy revolution requires major domestic policy and institutional reforms.** Governments can adopt climate-smart domestic policies now to deploy existing low-carbon technologies while a global climate deal is negotiated. Energy efficiency contributes to more than half of the emission reductions between the Sustainable Energy Development (SED) and Reference (REF) scenarios, and is fully justified by development benefits and future energy savings. To fully realize the huge energy efficiency potentials in the region requires the removal of fossil-fuel subsidies and incorporation of environmental externalities into energy pricing as well as a concerted strategy to tackle market failures and barriers with effective regulations, financial incentives, institutional reforms, and financing mechanisms. Under the SED scenario, low-carbon fuels for power generation—renewable energy and nuclear power—would meet half of the power demand by 2030. Scaling up renewable energy requires putting a price on carbon and providing financial incentives

1

to deploy renewable energy technologies. Not-yet-proven advanced technologies, such as carbon capture and storage, also are needed to bend the emission curve beyond 2030, but require accelerated research, development, and demonstration today.

- **Developed countries need to transfer substantial financing and low-carbon technologies.** To achieve this sustainable energy path, a major hurdle is to mobilize financing for the net additional investment of $80 billion per year over the next two decades. It is estimated that approximately $25 billion per year would be required as concessional financing to cover the incremental costs and risks of energy efficiency and renewable energy. In addition, substantial grants are also needed to build capacity of local stakeholders. The technical and policy means exist for such transformations, but only strong political will and unprecedented international cooperation will make them happen.

- **The World Bank Group is committed to scale up policy advice, knowledge sharing, and financing in sustainable energy to help the region's governments make such a shift.** The World Bank needs to increase efforts and focus future Bank energy business in East Asia on energy efficiency, renewable energy, and new technologies. Better integration of new and existing financing sources (IBRD, IDA, GEF, CTF, and carbon financing) can increase the magnitude and speed of the shift to a sustainable energy path.

- **The current report does not cover energy access issues in the region, nor propose any CO_2 emission targets.** This report outlines the strategic direction of the energy sector to meet its growing energy demand in an environmentally sustainable manner over the next two decades, and presents a pathway of policy frameworks and financing mechanisms to get there. A separate report will address the energy access issues in low-income countries—Cambodia, Lao PDR, Mongolia, Papua New Guinea, the Pacific Islands, and Timor-Leste—as well as part of Indonesia and the Philippines archipelago.

Twin Energy Challenges: Environmental Sustainability and Energy Security

For the last 3 decades, the East Asia region has experienced the fastest economic growth in the world, with a 10-fold increase in GDP.[1] The region has been less affected than other parts of the world by the ongoing global financial crisis. The high growth has been

1. The East Asia and Pacific (EAP) region within the World Bank Group's (WBG) operations comprises 12 diverse countries. This report covers only 6 of them and categorizes them into 2 groups: (1) China, and (2) EAP5, which comprises the 5 major economies of Southeast Asia: Indonesia, Malaysia, the Philippines, Thailand, and Vietnam.

accompanied by rapid urbanization, which has placed greater demands on urban energy services. The urban population is projected to increase by 50 percent over the next 2 decades. Energy consumption has more than tripled over the past 3 decades and is expected to double over the next 2 decades.

The remarkable growth and rapid urbanization have led to twin energy challenges in the region: environmental sustainability and energy security. East Asia has many of the world's most polluted cities. In addition, the region's carbon dioxide (CO_2) emissions have more than tripled over the past 20 years. The region includes some of the world's top greenhouse gas (GHG) emitters, although emissions per capita and historical emissions are still low compared to the developed countries. In addition, most East Asian countries are facing growing energy security concerns due to their increasing dependence on oil and gas imports.

Energy Scenarios: Reference and Sustainable Energy Path

The underlying study for this report examined 2 scenarios to 2030 to analyze how the region could better balance 3 competing objectives: (1) sustaining economic growth, (2) improving the local and global environment, and (3) enhancing energy security. These scenarios are not forecasts or energy plans. Their purpose is to help policymakers to gain a better understanding of the quantitative impacts of policy options. The reference scenario features a continuation of current government policies (REF scenario). REF does not mean business as usual. Rather, it assumes that governments' plans and targets will be realized in most countries but does not judge the likelihood of achieving them.[2] The alternative scenario aims for sustainable energy development (SED scenario). This scenario represents a low-carbon growth path with the lowest abatement costs under the constraints of the maximum technically feasible potential of energy efficiency and low-carbon technologies.[3] Finally, this study also conducted sensitivity analyses: (1) What if energy efficiency improvement were less than what is assumed in the SED scenario? and (2) What if new technologies could reach sizable scale prior to what is assumed in the SED scenario?[4]

---

2. This may underestimate the REF emissions, since some of the government's clean energy targets may not be achieved in reality.

3. In China, the REF scenario is based on historical energy efficiency efforts and government targets of renewable energy (RE) and nuclear power announced in 2005. The SED scenario extrapolates the most recent government targets of EE, RE, and nuclear power to 2030.

4. The SED scenario considers mostly proven technologies and excludes carbon capture and storage (CCS). While CCS is critically important in the long term beyond 2030, particularly for China, it is expected to be commercially available only after 2025.

Figure 1. Emission Gap between REF and SED Is Large, but Can Be Bridged by Energy Conservation and Low-Carbon Technologies, 2009–30 *(Gt)*

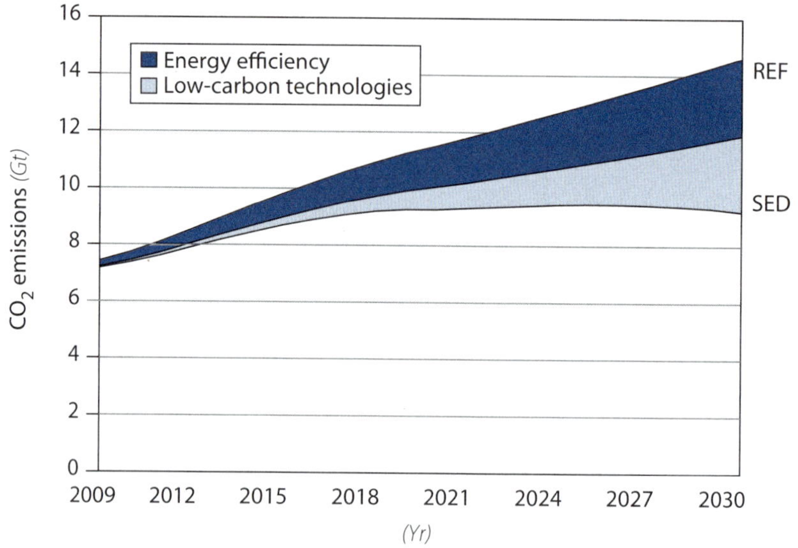

Source: Authors' calculations.

Baseline: Unacceptable Environmental Damages and Growing Concern with Energy Vulnerability

Under the REF scenario, emissions of local air pollutants and CO_2 will double over the next two decades (figure 1). Coal will continue to be the dominant fuel (figure 2). China continues to rely heavily on coal, whereas the share of coal slightly declines. The EAP5 countries plan to significantly expand the role of coal, as coal is abundant in the region and provides low-cost and secure energy supplies. The expansion of coal use will increase local air pollution and acid rain and exacerbate climate change.

Sustaining economic growth without compromising the environment is the greatest energy challenge facing East Asia over the next two decades. The EAP region is among the most vulnerable in the world to climate change threats. Particularly vulnerable are the large numbers of people living along the coasts and on low-lying islands (World Bank 2009a). Crop yields in many East Asian countries are projected to decline, due partly to rising temperatures and partly to extreme weather events (IPCC 2007). The 2006 Stern Review estimated that, without immediate action, the overall costs and risks of climate change could be at least 5 percent of global GDP each year and could cost more for vulnerable EAP countries (Stern 2006).

Figure 2. Power Generation Will Need to Shift Dramatically from Coal to Renewable Energy and Nuclear Power, 2007–30 *(%)*

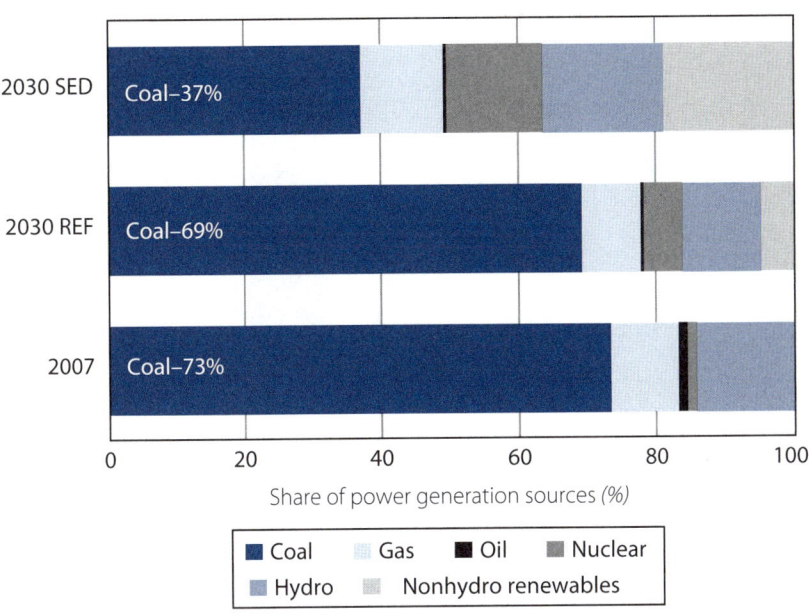

Share of power generation sources *(%)*

Legend: ■ Coal ▨ Gas ■ Oil ■ Nuclear ■ Hydro ▨ Nonhydro renewables

Source: Authors' calculations.

Energy security concerns will be heightened over increased risks of price volatility and exposure to disruptions in energy supplies. Throughout the next two decades, imports of oil and gas will grow across the region. Under the REF scenario, by 2030 China is expected to import 75 percent of its oil and 50 percent of its gas demand and become the largest oil importer in the world. Malaysia and Vietnam are projected to switch from being net energy exporters to net importers. By 2030, Thailand and the Philippines are expected to import 60 percent–70 percent of their energy needs.

Sustainable Energy Path: Improved Environment and Enhanced Energy Security

This study found that it is technically and economically feasible to stabilize CO_2 emissions by 2025 in East Asia provided that there are political will, institutional capacity, and transfer of financing and technologies from developed countries. In fact, such a low-carbon path produces substantial benefits for economic development through energy savings, better public health, enhanced energy security, and job creation.

Under the SED scenario, CO_2 emissions of China and EAP5 countries could peak at 2025 and decline slightly thereafter. By 2030,

Figure 3. SED Will Improve Local Environment and Energy Security

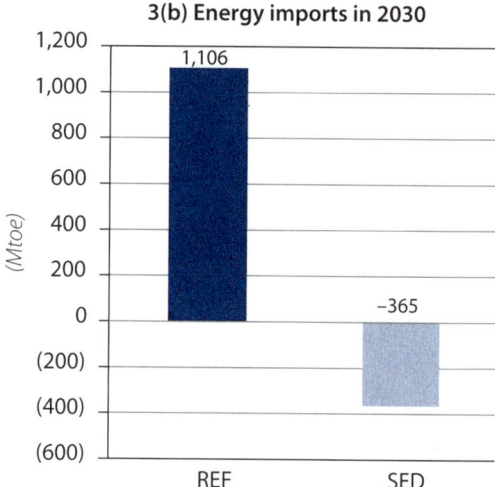

Source: Authors' calculations.

CO_2 emissions would reach 9.2 Gt, 37 percent below the REF scenario (figure 1). China is the main source of these emission reductions, consistent with its 80 percent share of the region's energy consumption and 85 percent of the region's CO_2 emissions. These projected results are comparable to those of other leading international and Chinese studies.[5] Local environment also will be substantially improved, with a 50 percent reduction in damage costs by 2030 under the SED scenario compared to the REF scenario (figure 3a).

Energy security also will be improved by increasing fuel diversity and reducing imports (figure 3b). Renewable energy can hedge against fossil fuel price volatility. For example, increasing fuel prices by 20 percent increases generation costs by 16 percent for gas and 6 percent for coal, but leaves renewable energy practically untouched (WEF 2009). In 2030 the SED scenario would reduce annual Chinese oil imports by more than 240 million tons of oil equivalent (Mtoe), or a 38 percent cut, compared to the REF scenario. By the same year, Malaysia and Vietnam would switch from being net energy importers under the REF scenario to net exporters under the SED scenario.

However, improving environmental sustainability and enhancing energy security have trade-offs. For example, the SED scenario projects more gas consumption in gas-producing countries and a

5. McKinsey Global Institute 2008; China Energy Research Institute 2009; Clarke and others 2009; and Riahi and others 2007; IIASA 2009.

higher share of nuclear power in the generation mix in China. An increased role of gas and nuclear power could increase risks of gas price volatility, nuclear reactor safety, and nuclear weapons proliferation. In addition, a relatively large share of intermittent renewable energy (such as solar and wind) in the electricity grids may affect reliability of power supply. Unreliability can be addressed in various ways: hydropower (including pumped storage), load management, energy storage facilities, interconnection with neighboring provinces and countries, and smart grid management.

Paradigm Shift: Use Energy Wisely and Embrace Low-Carbon Technologies

The sustainable energy path requires paradigm shifts to a new low-carbon development model and lifestyle. Developing countries in East Asia will need to avoid the carbon-intensive path and wasteful lifestyles followed by developed countries in the past. Developing countries can pursue sustainable lifestyles by promoting novel low-carbon urbanization models that focus on compact city design, enhanced public transport, green buildings, clean vehicles, and distributed generation.

Greater energy efficiency improvements and increasing penetration of low-carbon technologies both are needed to achieve a sustainable energy future. In the short term, the first step is to increase energy efficiency. Energy conservation contributes to more than half of the emission reductions from the SED to the REF scenario in 2030 (figure 1) and is fully justified by development benefits and future energy savings. In the short to medium term, the second largest source of emission reductions comes from low-carbon technologies for power generation, particularly renewable energy. Over the long term (beyond 2030), not-yet-proven advanced technologies are critical to bend the emission curve. Historically, innovation and technology breakthroughs have reduced the costs of overcoming formidable technical barriers, given effective and timely policy action—a key challenge facing the world today.

To realize the region's huge energy efficiency potential requires major policy and institutional reforms, adjustment in economic structure, technological innovations, and changes in consumer behavior. Under the SED scenario, China will need to reduce energy intensity by 4.3 percent annually over the next 2 decades. The past record showed a reduction of 3.4 percent per year over the last decade. Currently, the government is targeting a reduction of 4.2 percent per year. This is a daunting goal, given that China is at a

Table 1. Sensitivity Analyses Show That Energy Efficiency Is the Most Important Emission Reduction Option and That New Technologies Can Further Reduce Emissions

	Peaking year	Emissions in 2030 (Gt)	Financial costs (US$bil/yr)	Investment cost (US$bil/yr)
REF scenario	—	14.6	1,167	100
SED scenario	2025	9.2	1,081	180
SED + new technologies scenario	2021	8.7	1,096	195
Low energy efficiency scenario	—	11.1	1,129	160
Low energy efficiency + new technologies scenario	2027	10.3	1,141	170

Source: Authors' calculations.
Note: Financial costs include capital investment, operations and maintenance (O&M), and fuel costs.

developmental stage in which energy-intensive industries, driven by demand from domestic and export production, dominate the economy. Economic structural change toward a less energy-intensive economy could be the single largest contributing factor to such a reduction. Under the SED scenario, other East Asian countries also would need to substantially reduce their energy intensities—3.1 percent per year for Indonesia and the Philippines until 2030.

While these assumptions are technically feasible, this study conducted sensitivity analyses to test their robustness. If the energy efficiency improvements were half-way between the REF and SED scenarios, East Asia's CO_2 emissions would peak by 2030 (a 5-year delay in peaking time compared to the SED scenario)—increasing by 20 percent that year with a total additional financial cost of 4 percent (table 1). This sensitivity analysis demonstrates the critical role of energy efficiency measures.

Under the SED scenario, low-carbon technologies would meet half of East Asia's power demand by 2030. This achievement would require a 3-fold increase in the share of low-carbon technologies (renewable energy and nuclear power) in power generation from today's 17 percent (figure 2). This large increase would come primarily from hydropower, wind, and biomass in China; hydro, biomass, and geothermal in Indonesia; geothermal and hydro in the Philippines; imported hydro and biomass in Thailand; and hydro in Vietnam. The expansion of nuclear power from the REF to the SED scenario is projected to come solely from China, because of the government's aggressive plans to boost nuclear power. Under the REF scenario, the governments of Malaysia, Thailand and Vietnam plan to introduce nuclear power after 2020.

The share of coal in power generation is projected to decline from 70 percent under the REF scenario to 36 percent under the SED scenario by 2030, and carbon capture and storage will play a key role.

The future role of coal in a carbon-constrained world will depend increasingly on widespread use of carbon capture and storage (CCS)—a promising technology yet to be proven on a large scale. Given the technological uncertainty, this study conducted a sensitivity analysis of aggressive deployment of new technologies, which would represent the upper boundary of deployment of CCS and solar energy. This analysis assumes that CCS technology could become commercially available on a large scale by 2020 and that the costs of solar energy could be significantly reduced.[6] As a result, CO_2 emissions would peak in 2021 (4 years earlier than the SED scenario) and would decline by 5 percent in 2030 (table 1), with a total additional financial cost of 1 percent, compared to the SED scenario. If energy efficiency improvements were half-way between the REF and SED scenarios but CCS and solar technologies were aggressively deployed, CO_2 emissions would peak in 2027 and increase by 12 percent (table 1), with a total additional financial cost of 6 percent by 2030, compared to the SED scenario. This sensitivity analysis demonstrates that energy efficiency is the most important and cost-effective emission reduction option and cannot be replaced by new technologies.

Under the SED scenario, natural gas is expected to play an important role in East Asia. Southeast Asian countries, including Indonesia, Malaysia, and Vietnam, have large gas reserves and resources. With favorable policies and regulations, gas production and consumption could increase substantially and play an important role in meeting the region's energy demand over the next two decades. However, in China, with the current gas share of 3 percent of primary energy supply, the role of gas will continue to be limited in 2030.

Financing the Sustainable Energy Path: Affordable but Requires International Concessional Financing

A formidable challenge is to mobilize financing for the net additional average investment of $80 billion per year to achieve this sustainable energy path over the next two decades.[7] The financing needs are estimated at $85 billion per year in energy efficiency for the power, industry, and transport sectors; and $35 billion per year for low-carbon technologies ($25 billion for renewable energy and $10 billion for nuclear power) (figure 4). In the mean time, due to energy efficiency measures, the SED scenario avoids an average of $40 billion per year in investments in thermal power plants. This savings results

6. It is assumed that CCS would account for 25% and 10% of the coal-fired power capacity in China and each of the EAP5 countries, respectively, by 2030.
7. The additional investment needs range from $45 billion in 2010 to $255 billion in 2030.

Figure 4. Sustainable Energy Path Requires Substantial Additional Financing *(US$bil)*

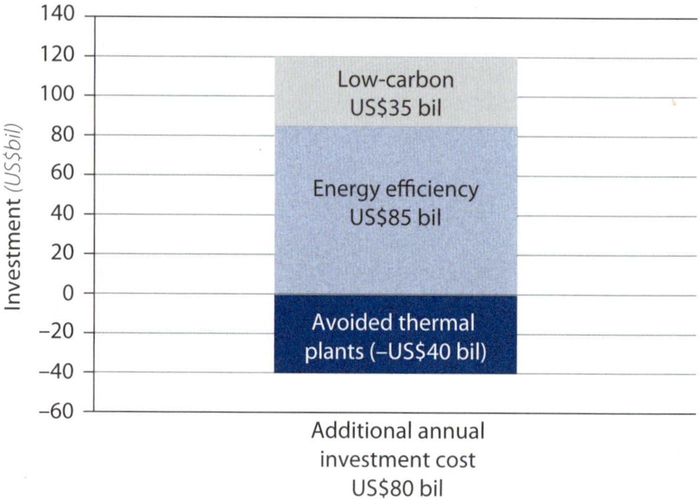

Source: Authors' calculations..

in a net additional average capital investment of $80 billion per year from now to 2030, or an average 0.8 percent of regional GDP.

The energy savings largely will pay for this additional investment cost. While energy efficiency, many renewable energy technologies, and nuclear power require additional upfront investment, they can recoup part or all of that investment through lower energy spending from fuel savings in future years. Therefore, the total financial costs of the SED scenario (capital investment, operations and maintenance, and fuel costs) would not exceed those of the REF scenario after the first three years.

However, the upfront financing requirements will be a major hurdle. Historically, financing has been a constraint in developing countries. The results have been under-investment in infrastructure and a bias toward energy choices with lower upfront capital costs, even when such choices were known to eventually result in higher overall costs. Many clean energy investments have high upfront capital costs, followed by savings in fuel costs in future years. In fiscally constrained developing countries, these high upfront capital costs are a significant barrier to investments in low-carbon technologies. To overcome this barrier, concessional financing is required to cover the incremental costs and risks of clean energy solutions.

Financing Energy Efficiency

Financing the upfront investments in energy efficiency faces major challenges, despite their lifetime "negative" costs (fuel savings that are greater than additional investments). Many energy-efficiency measures are financially viable for investors at current prices but have not been fully realized due to many market failures and barriers. Energy efficiency investments tend to be small and have high transactions costs. They also are perceived as risky since investors are not sure whether the expected future savings will be realized. As yet, financial institutions lack the required expertise, interest, and incentives to develop the energy efficiency business line, compared to traditional supply-side investment options. As a result, domestic banks in the region do very little energy efficiency lending. While lack of domestic capital is rarely the problem, inadequate policy frameworks and institutional capacity are significant constraints to financing. Tackling these barriers requires interventions, but they, too, entail additional costs.

Concessional financing, along with technical assistance grants, is required to cover the incremental risks, financial incentives, and transaction costs. This study estimated that the required concessional financing could amount to approximately 20 percent of the projected $85 billion per year (or $17 billion) of additional investment needs in energy efficiency for mitigation of perceived risks and financial incentives.[8] This estimate varies by sector. The small and medium-sized enterprises require more risk mitigation, as their perceived risks are higher than those of large-scale enterprises. Individual consumers in the residential sector usually demand very short payback times, and financial incentives often are required to offset the higher upfront costs of energy-efficient products. More than one-third of the $85 billion additional financing would be invested in the transport sector, which also requires financial incentives for consumers to shift to more efficient vehicles.[9] The need for such concessional financing is expected to decline over time. The remainder is likely to be met by commercial public and private financing. In addition to the $85 billion investment needs, substantial grants for building capacity of local stakeholders will be critical, particularly for countries with lower institutional capacity and market penetration of energy efficiency (figure 5).

Three main financing mechanisms have been developed for energy-efficiency investments. They are (1) loans and partial loan guarantee schemes operating within commercial banks or as specialized agencies or revolving funds; (2) energy efficiency and demand-management funds, financed by a surcharge on electricity consumption

---

8. See chapter 2 for detailed methodology and assumptions.
9. The investments in the transport sector include capital costs of only efficient vehicles, not those of public transport infrastructure. Therefore, they underestimate the investment needs in the transport sector.

Figure 5. Concessional Financing Is Critical

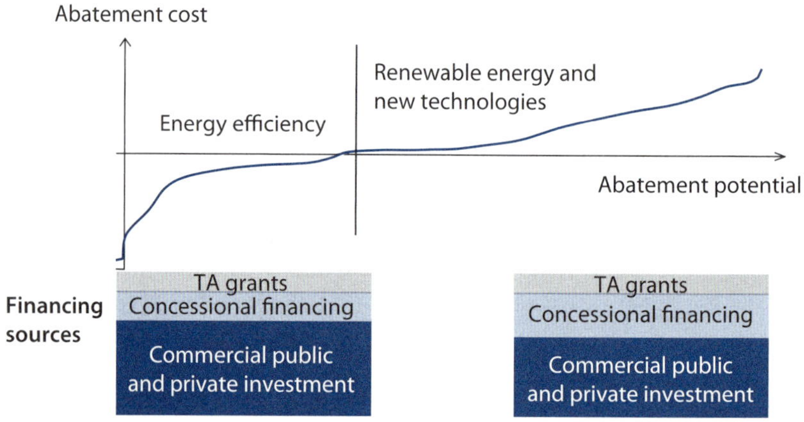

Source: Authors' calculations.

(system benefit charge) or a government budget, and managed by utilities or dedicated agencies; and (3) third-party financing through energy service companies (ESCOs) (Taylor and others 2008).[10]

Financing Renewable Energy

Many renewable energy technologies are economically but not financially viable (good for countries but not sufficiently profitable for investors). The costs of renewable energy have dropped dramatically over the past two decades and are expected to decline rapidly along the learning curve in the near future. With rising fossil-fuel prices, the cost gap is closing.

Concessional financing, along with technical assistance grants, is required to cover the incremental costs of renewable energy (figure 5). This study estimated that the required concessional financing could amount to approximately one-third of the projected $25 billion per year (or $8 billion) of additional investment needs in renewable energy to cover the incremental costs. The reminder is likely to be met by commercial public and private financing, comprising $9 billion baseline financing[11] and $8 billion from avoided fossil fuel costs that would be recovered by fuel savings. Substantial grants for capacity building also will be critical.

In reality, the amount of required concessional financing will be determined by the renewable energy tariff of each country. If the tariff is set high enough to ensure the financial viability of renewable energy projects, there will be no need for concessional financing. In

---

10. ESCOs provide energy-efficiency services and financing to clients.
11. The investment costs of coal-fired power plants to deliver the same amount of electricity generation.

such a case, domestic consumers pay for the incremental costs between renewable energy and fossil fuels. As the share of renewable energy in the power mix increases, the total incremental costs in the whole power system will rise from minimal to relatively high, which could increase financial burdens on domestic consumers, particularly on the poor. Carbon financing can improve revenue streams but will be unlikely to meet the bulk of this financing need. Initial experience with the Clean Development Mechanism (CDM) showed that, in its best year, approximately $1 billion of new CDM projects were registered—only 1 percent of the projected net of $80 billion financing needs (Capoor and Ambrosi 2009). Even with CDM reforms, carbon financing likely will remain a small part of the solution. Therefore, additional concessional financing, such as a scaled-up Clean Technology Fund, will be needed to bridge the financing gap.

Achieving a Sustainable Energy Path: Policy Tools for Transformations

Governments need to take immediate action to transform the energy sector into one of high energy efficiency and widespread diffusion of low-carbon technologies. The long lives of energy capital stocks (power plants, buildings, roads) mean that investments over the next decade largely will determine emissions through 2050. Delaying action will lock the region into a high-carbon infrastructure, requiring future more costly retrofitting and premature scrapping of existing capital stocks. However, the inertia also offers a large opportunity to build efficient low-carbon technologies into new infrastructure, as half of the energy stocks needed in developing countries by 2020 have yet to be built (McKinsey Global Institute 2009a).

Strong government commitment and political will are essential to unlock the opportunities of sustainable energy. Many East Asian countries are taking steps in the direction of sustainable energy. However, to get onto this path, accelerating the speed and scaling up the efforts are needed. A paradigm shift is required to leapfrog to new development models and lifestyle changes toward energy conservation and sustainability. Policy tools exist for large-scale deployment of energy efficiency and low-carbon technologies, but these tools need to be tailored to the maturity and costs of technologies and national context (figure 6).

Tapping the Huge Potential of Energy Efficiency

To realize the huge potential of energy efficiency in the region requires energy pricing/taxing reforms, coupled with regulations, incentives, and institutional reforms. In the near term, energy

Figure 6. Policy Tools Need to Be Tailored to Maturity and Costs of Technologies

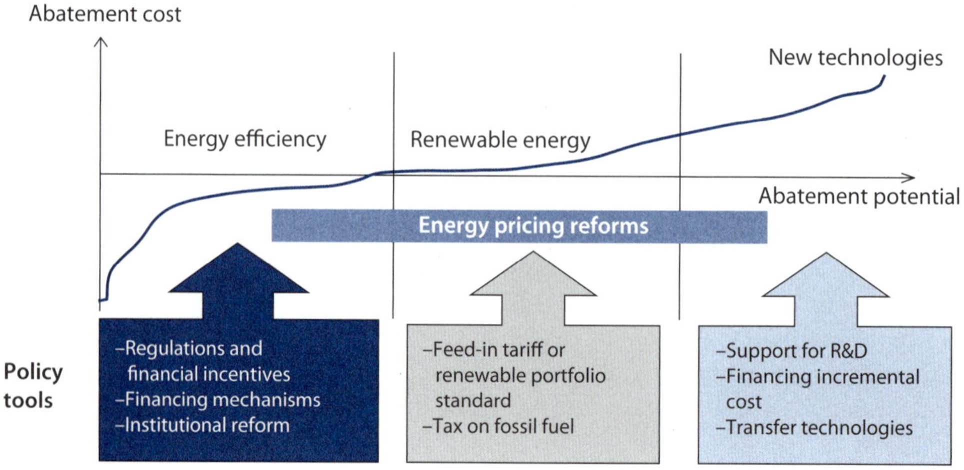

Source: Authors.

efficiency is the largest and lowest cost source of emission reductions. Small-scale, fragmented energy-efficiency measures that involve multiple stakeholders and tens of millions of individual decisionmakers are more complex organizationally than large-scale, supply-side options.

Pricing policies. Market-based pricing reforms are fundamental to an efficient, sustainable, and secure energy sector (Berrah and others 2007). Price is a driving force to stimulate energy efficiency improvements, discourage energy waste, mitigate rebound effects, and encourage clean energy technologies. Energy prices should (1) remove fossil fuel subsidies; (2) internalize environmental costs through appropriate use of a fuel tax and/or a carbon tax; and (3) provide incentives to invest in end-use energy efficiencies such as investment subsides, soft loans, consumer rebates, and tax credits. In 2007 East Asia's fossil fuel subsidies ($70 billion) (IEA 2008a) were close to the estimated additional net financing required for a sustainable energy path ($80 billion). Fuel taxes have proven to be one of the most cost-effective ways to reduce transport energy demand. In reality, however, increases in energy prices are no simple matter. They require strong political will and effective social protection targeted at low-income groups.[12]

Regulations. Regulation is one of the most cost-effective means to improve energy efficiency. Economy-wide energy-intensity targets,

---

12. An energy price increase is often met with stiff political resistance. For example, adjusting electricity tariffs or gasoline prices before elections becomes a highly contentious issue.

appliance standards, building codes, industry performance targets (energy consumption per unit of output), and fuel-efficiency standards are among the most cost-effective measures. However, weak enforcement of regulations is a problem in many East Asian countries. Therefore, pricing and fiscal policies should go hand in hand with regulations.

Institutional reforms. Given the fragmented nature of energy efficiency measures, a national institutional champion is essential. For example, a dedicated energy-efficiency agency can play an important role in coordinating multiple stakeholders, implementing energy-efficiency programs, and raising public awareness. However, this agency requires adequate resources, the ability to engage multiple stakeholders, autonomous decisionmaking, and credible monitoring of results (ESMAP 2008).

Market-based mechanisms such as ESCOs are important to implement energy efficiency by delivering technical and financing services. To strengthen ESCOs and mainstream their business models require policies, financing, and technical support. In China, for example, after a decade of capacity building supported by the WB/ GEF and the government, the ESCO industry grew from 3 companies in 1997 to more than 400. Chinese ESCOs had $1 billion in energy performance contracts in 2007 (World Bank 2008a).

Scaling up Renewable Energy

Scaling up renewable energy requires financial incentives and/or an energy tax or a carbon tax to internalize local and global environment externalities. The two most effective financial incentive policies to mandate and scale up renewable power generation are feed-in tariffs and renewable energy portfolio standards (RPS). Feed-in tariffs require mandatory purchases of renewable energy at a fixed price, while RPS requires utilities in a given region to meet a minimum share of power or level of power generation from renewable energy (ESMAP 2006). Additional financial incentives include reducing capital and operating costs through investment or production tax credits; improving revenue streams with carbon credits; and providing financial support through concessional loans and guarantees (World Bank 2006b).

Accelerating Innovation and New Technologies

Deploying advanced technologies on a large scale requires enhanced research, development, and demonstration (RD&D) today, coupled with an adequate carbon price. Proven technologies can meet the bulk of the abatement needed in the short to medium term. However, innovations and new technologies (Brown and others 2005) are critical to bend the emission curve over the long-term beyond 2030. Given the long lead time needed to develop technology, now is the time for East Asian nations to ramp up RD&D.

Developed countries should take the lead, but developing countries cannot afford to wait. The clean technology revolution offers an opportunity for developing countries to leapfrog to the next generation of new technologies, create local manufacturing industries to drive down costs, and become global technology leaders. These benefits already are among the main drivers for the strong and rapid R&D in some emerging economies. The largest barrier is the high incremental cost between low-carbon technologies and conventional energy options, particularly in developing countries. Effective, innovative, fair, and affordable methods are needed to accelerate the transfer of low-carbon technologies and the financing of incremental costs to developing countries.

Transforming Urban Forms to Low-Carbon Cities

The speed and scale of the urbanization trend in the region presents an unrivalled opportunity to build low-carbon cities today with compact urban designs, public transport, green buildings, clean vehicles, and distributed generation. Currently, East Asia's GHG emissions are dominated by the power and industrial sectors. However, over the next two decades, the transport and building sectors are expected to grow faster than the power and industrial sectors, due to unprecedented urbanization. Smart urban planning that avoids urban sprawl—higher density, more spatially compact, and more mixed-use urban design that allows growth near city centers and transit corridors—can substantially reduce energy demand and CO_2 emissions (World Bank 2009a). In addition, in new buildings, even existing energy efficiency technologies, can save up to 40 percent of energy use cost-effectively. Integrated zero-emission building designs combine energy-efficiency measures with on-site power and heat generation from solar and biomass. They are technically and economically feasible—and their costs are falling (Brown and others 2005). This study also found that transport energy use and emissions in East Asia can be reduced by 38 percent from the baseline by 2020. These reductions can be achieved via more efficient vehicles to meet EU fuel economy standards, coupled with energy-smart urban planning, public transport, and pricing policies. Modal shifts to mass transit also have large development co-benefits of time-savings in traffic, less congestion, and better public health from reduced local air pollution.

World Bank Group's Role: Accelerate Shift toward Sustainable Energy

The World Bank Group has supported energy efficiency and renewable energy in EAP, but a substantial scale-up is required. Energy efficiency and renewable energy accounted for 40 percent of the Bank's EAP energy portfolio over the past decade. However, key findings of this study show that to put the energy sector on a sustainable path, a substantial scale-up in both lending and analytical and advisory activities (AAA) in sustainable energy will be required across the World Bank Group.

The future WBG operational energy strategy in EAP will focus on four main areas: (1) energy efficiency, (2) renewable energy, (3) new technologies for sustainable energy, and (4) energy access expansion. These pillars are consistent with the twin objectives in the Bank's proposed Energy Strategy: (1) improving access and reliability of energy supply; and (2) facilitating the shift to a more environmentally sustainable energy development path. Furthermore, the EAP infrastructure sectors can achieve considerable synergies through better cooperation, for example, in building low-carbon cities with a three-legged approach: transforming buildings and vehicles to be more efficient (including electric vehicles), transforming mobility toward mass transit, and transforming urban planning.

The WBG is well positioned to provide policy advice, facilitate knowledge sharing, and catalyze financing to help the governments in the region shift to a sustainable energy path. To achieve the SED scenario requires policy and institutional reforms and concessional financing. The Bank will step up policy advice, institutional strengthening, and knowledge sharing in sustainable energy. Finally, closer coordination with other international financial institutions (IFIs) and development partners will be needed to maximize the effectiveness of policy advice and provide catalytic financing to narrow the gap of the additional concessional financing needed to make this shift.

Better integration of new and existing financing sources (IBRD, IDA, GEF, the Clean Technology Fund or CTF, and carbon financing) can increase the magnitude and speed of the shift to a sustainable energy path. Grants from GEF and other donors are used to set up an enabling environment, build capacity, and share transaction costs and risks associated with energy efficiency and renewable energy. IBRD/IDA funds are used to provide long-term financing for the capital costs of low-carbon projects. CTF funds provide concessional financing to buy down the incremental costs and risks of low-carbon technologies. Carbon financing adds an additional revenue stream to improve the financial viability of sustainable energy projects.

The regional or subregional programs have the advantages of facilitating knowledge-sharing across countries and promoting regional trade. Such programs can send the right signals that the Bank is committed to supporting sustainable energy in the region. In addition, the programs can help create a large regional clean energy market attractive to private investors and entrepreneurs. The Bank can facilitate dissemination of the region's successful experience and introduce international best practices. In addition, regional hydropower trade can provide the least-cost energy supply with zero carbon emissions to countries in Southeast Asia.

The immediate follow-up to this study will be country-level policy dialogues and AAA to determine the strategic direction and identify the opportunities of each country program. Given its sheer size and the government's commitment, China leads the WBG's clean energy portfolio of energy efficiency and renewable energy. In other EAP countries, country-specific AAA can help pinpoint energy efficiency opportunities and the economically viable potentials of each renewable energy resource, identify barriers, and propose interventions.

Energy Efficiency

Energy efficiency improvements include both supply and demand. All EAP countries have huge potential to improve energy efficiency on the demand side in the industrial, building, and transport sectors. They also have large potential to improve efficiency on the supply side in the power sector, such as rehabilitating coal-fired power plants, fuel switching from coal to gas, and reducing transmission and distribution losses. Supply-side energy efficiency has been mainstreamed into the Bank's operations. However, the Bank Group also should strengthen and scale up demand-side energy efficiency, as its potential is much larger (IEA 2008b).

The WBG can help EAP countries develop and implement policy and institutional reforms, financing mechanisms, and market-based delivery mechanisms to scale up energy efficiency. A decade of WBG experience in energy efficiency demonstrates that these efforts must shift from developing technologies to delivering services and savings. In addition, both regulations and financial incentives are required to transform energy efficiency markets. The appropriate balance between the two varies from country to country. The major constraints to energy efficiency are inherently institutional. Scaling up energy efficiency requires a successful institutional framework, technical and management capacity, and strong coordination and cooperation among governments at every level (World Bank 2009b).

Financial intermediary and risk mitigation programs have proven successful in mainstreaming energy efficiency financing within the domestic banking sector. In East Asia, local financial institutions have had little experience in financing energy efficiency so are unlikely to enter this line of business without external support. The WBG can play an important role in helping domestic financial institutions to increase their confidence and capacity in assessing energy efficiency projects through risk guarantee instruments, the financial intermediary approach, and capacity building. The Bank's successful experience in China could be replicated in other East Asian countries. A key lesson learned is the importance of technical assistance (TA), particularly at the beginning, to provide training and advisory services to local banks to develop financial structures and to build the capacity of project developers (Taylor and others 2008).

The energy efficiency investments can focus initially on sectors with high energy savings potential, such as large industrial customers and public/commercial buildings. Industry dominates energy consumption and has the largest energy saving potential in most East Asian countries. The Bank's energy efficiency programs can focus on energy-intensive equipment (for example, boilers and motors) and/or energy-intensive subsectors (for example, iron and steel, cement, and chemical industries) through TA on efficiency standards, sector performance targets, and industry energy management standards, combined with investment in energy efficiency implementation. Buildings represent the second largest energy saving potential in many East Asian countries. The Bank can target commercial and public buildings through TA on policy interventions, business models, and zero-emission building technologies, combined with investments.

Renewable Energy

Scaling up grid-connected renewable energy requires enabling legal, policy, and regulatory frameworks and long-term financing. Lessons learned from a decade of WBG experience in grid-connected renewable energy demonstrate three key successful factors: sufficient and stable pricing through long-term power purchase agreements, mandatory purchase by utilities, and passing through incremental costs to consumers (World Bank 2006b). The WBG can provide grants to help develop and implement such legal, policy, and regulatory frameworks; fund pre-investment activities, resource assessment, and capacity building; and offer long-term financing to capital-intensive renewable energy projects. The successes of China and Vietnam could be replicated in other East Asian countries.

New Technologies

The WBG can provide cost-shared R&D and facilitate technology transfer (TT) to help EAP countries leapfrog to the next generation of clean energy technologies. The Bank can tap GEF and other grants to aid in purchasing licenses, and provide cost-shared grants to support applied R&D funding toward promising clean technologies and/or adoption of international quality standards. The projects of the World Bank in China have helped boost the country's solar-photovoltaic (PV) and wind-power manufacturing industries. In addition, a guaranteed market approach (an incentive scheme to provide a large guaranteed market to companies that develop and diffuse breakthrough technologies) could substantially reduce costs of technology development through economies of scale. Finally, few existing financing sources pay for the high incremental costs of new technologies; hence, additional financing mechanisms are needed. For example, even though CCS is critically important for China's coal-based economy under a carbon constrained future, CCS is not eligible for CTF.

Cross-Cutting Solutions: Policy Advice, Institutional Strengthening, and a Sustainable Energy Knowledge Hub

To accelerate the shift to a sustainable energy path, the Bank needs to ramp up policy advice, institutional strengthening, and knowledge-sharing in the areas outlined above. The Bank's value-added in China and EAP5 countries lies in its advisory services and new ideas from international experience and best practices. The WBG needs to increase its efforts to advise country clients on putting in place an enabling environment conducive to scaling up energy efficiency and renewable energy, leveraging significant private sector investment, and increasing market penetration of low-carbon technologies. Institutional strengthening also is important, particularly innovative institutional models to manage and implement energy efficiency. For example, advisory services and knowledge sharing are critical to jumpstart energy efficiency lending in domestic financial institutions.

In this regard, this report proposes a World Bank Group East Asia Sustainable Energy Forum (EASEF). This forum would aim to engage countries on clean energy policy advice, regional energy cooperation and investments, knowledge sharing, capacity building, and technology development. The forum would promote the sustainability agenda, support harmonization of regional energy markets, provide advisory services, disseminate cutting-edge knowledge, and facilitate South-South cooperation on sustainable energy and climate change.

To raise the profile of sustainability and gain political support, this study also proposes that the Bank's EAP Region call for a high-level Regional Summit. With participation from EAP ministers, this summit would strive to put energy efficiency, renewable energy, new technologies, and climate change at the top of EAP nations' agenda and seek the ministers' support for the establishment of EASEF.

Regional Energy Challenges

Key messages: Over the past three decades, the East Asia region has experienced the world's fastest economic growth, accompanied by rapid urbanization. China relies heavily on coal, whereas oil and gas dominate the energy mix in EAP5 countries. This remarkable growth has led to twin energy challenges in the region: environmental sustainability and energy security. This study explores sustainable energy development paths in East Asia to better balance three competing objectives: sustaining economic growth, improving local and global environment, and enhancing energy security.

Defining the Study Region and Its Country Groupings

The East Asia and Pacific (EAP) Region within the World Bank Group's (WBG) operations comprises 12 diverse countries with unique energy challenges. This study covers only 6 of these countries and categorizes them into 2 groups:

1. China is treated separately because it alone accounts for 85 percent of regional energy consumption and CO_2 emissions. China relies heavily on coal.

2. EAP5 comprises the five major economies of Southeast Asia: Indonesia, Malaysia, the Philippines, Thailand, and Vietnam. Except for Vietnam, they are middle-income countries, and rely heavily on oil and gas. Indonesia, Malaysia, and Vietnam have rich gas resources. Thailand and the Philippines are net energy-importing countries.

This study selected only the above 6 countries because, as a result of high economic growth and rapid urbanization, they share the common key energy challenges of environmental sustainability and energy security.

Figure 1.1 EAP Is Categorized in Three Country Groupings

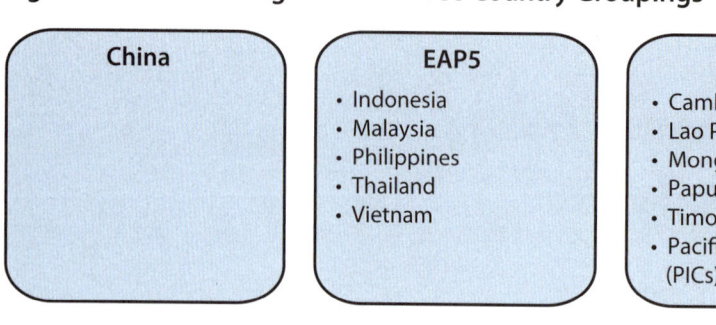

Source: Authors.

A separate companion report will focus on the remaining six countries of EAP 12-6: Cambodia, Lao PDR, Mongolia, Papua New Guinea, Timor-Leste, and the Pacific Island Countries (PICs). These countries are mostly low-income countries, and heavily rely on oil imports, except Mongolia, whose energy mix is dominated by coal. Energy access and fuel security are top priorities in these countries.

1.1 East Asian characteristics: High economic growth and rapid urbanization

High Growth Rates

The East Asia region has experienced the fastest economic growth in the world for the last three decades (figure 1.2), achieving a 10-fold increase in GDP. The highest growth rate came from China. The region has been less affected by the ongoing global financial crisis, and its economic growth is expected to continue in 2010.

Rapid Urbanization

East Asia's high economic growth has been accompanied by rapid urbanization, which generated greater energy demands for residential and transport services (figure 1.3). The world's cities already consume more than two-thirds of global energy and produce more than 70 percent of CO_2 emissions (IEA 2008a). In 2008, for the first time in history, more than half the world's population–3.3 billion people–lived in urban areas. By 2030, this number is expected to swell to almost 5 billion. The unprecedented scale of urban growth will be particularly notable in Asia, whose urban population is expected to increase by 50 percent between 2000 and 2030 (UNFPA 2007a). East Asia's urban share of its total population is expected to rise from the current 46 percent to 60 percent by 2030.

The current trend shows that many cities in the region grow through sprawl rather than densification. As a result, demand for

Figure 1.2 EAP Has Experienced World's Highest Economic Growth over Past 25 Years, 1980–2005 *(%)*

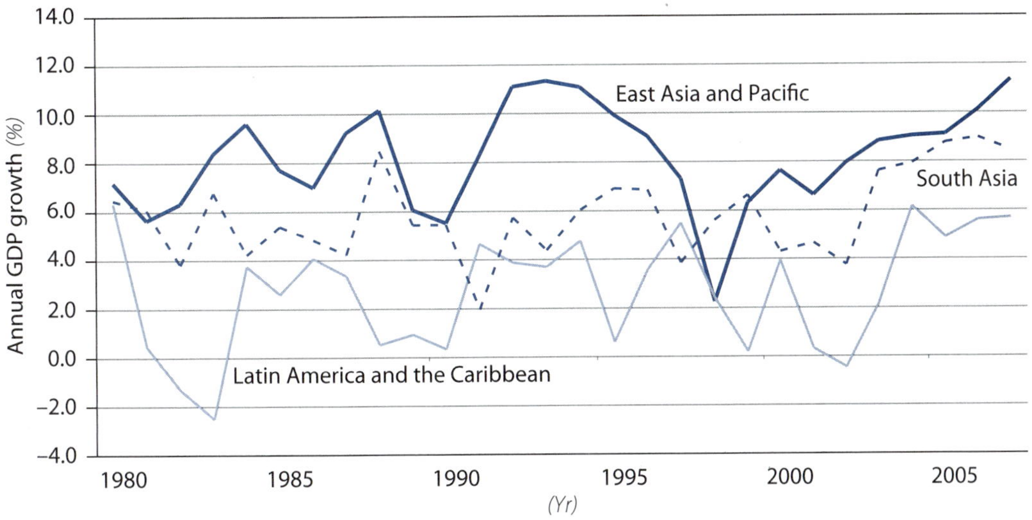

East Asia and Pacific

South Asia

Latin America and the Caribbean

Source: Authors based on data from World Bank 2009c.

travel will increase in ways not easily served by public transport. Low-density settlements also make it more difficult to adopt energy-efficient district heating for buildings.[14] In addition, public transport has not kept up with urban growth in many countries, so the move

Figure 1.3 Rapid Urbanization in East Asia, 1950–2050 *(%)*

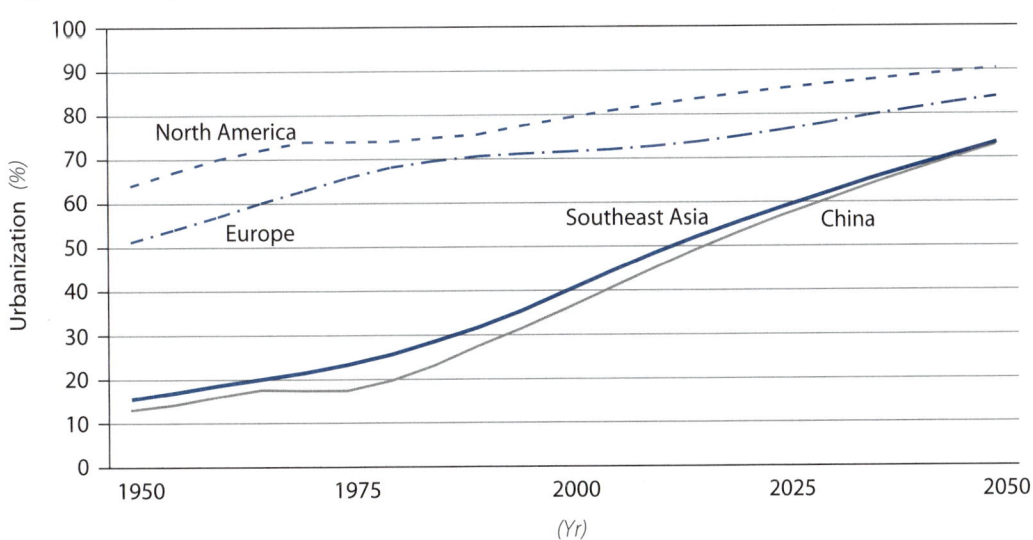

North America

Europe

Southeast Asia

China

Source: Authors based on data from United Nations 2007b.

---

14. District heating distributes heat generated in a centralized location for residential and commercial buildings, supplied by cogeneration plants or large-scale heating boilers, with higher efficiency.

to individual car ownership is causing chronic and increasing problems of congestion.

In many East Asian countries, a substantial proportion of the urban population is surviving through a precarious existence in the urban informal sector and is vulnerable to food and fuel price shocks. Unless the specific concerns of this constituency are addressed through targeted programs, energy reform measures may be dissipated by political opposition.

1.2 East Asia faces twin energy challenges: Environmental sustainability and energy security

Energy consumption increases with income per capita, population, and energy intensity (defined as energy consumption per unit of GDP). Energy intensity is determined by economic structure (manufacturing and mining are more energy intensive than agriculture and services), energy efficiency, and energy-consuming lifestyles. A seven-fold increase in economic growth combined with significant decline in energy intensity (coming mainly from China) has tripled energy consumption over the past 25 years (figure 1.4).

The decoupling of energy growth from economic growth is a remarkable achievement. China has made significant progress

Figure 1.4 EAP Energy Consumption Tripled in 25 Years, Driven by Economic Growth, 1980–2004 *(Mtoe)*

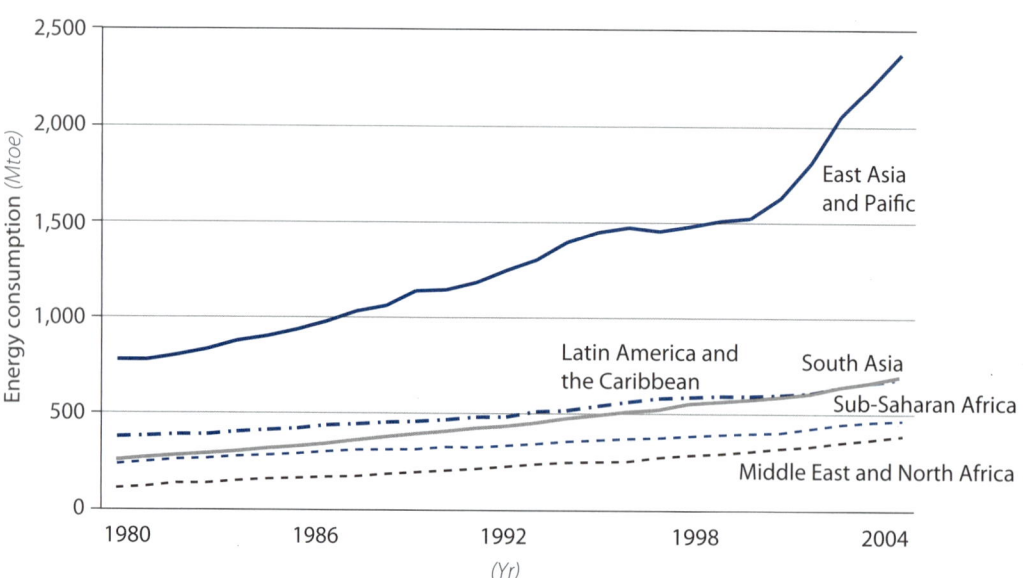

Source: Authors based on data from World Bank 2009c.

Figure 1.5 China and Vietnam Significantly Reduced Energy Intensity Whereas Other East Asian Countries Increased It, 1980–2006, but Region Lags behind Developed Countries *(toe/M$GDP)*

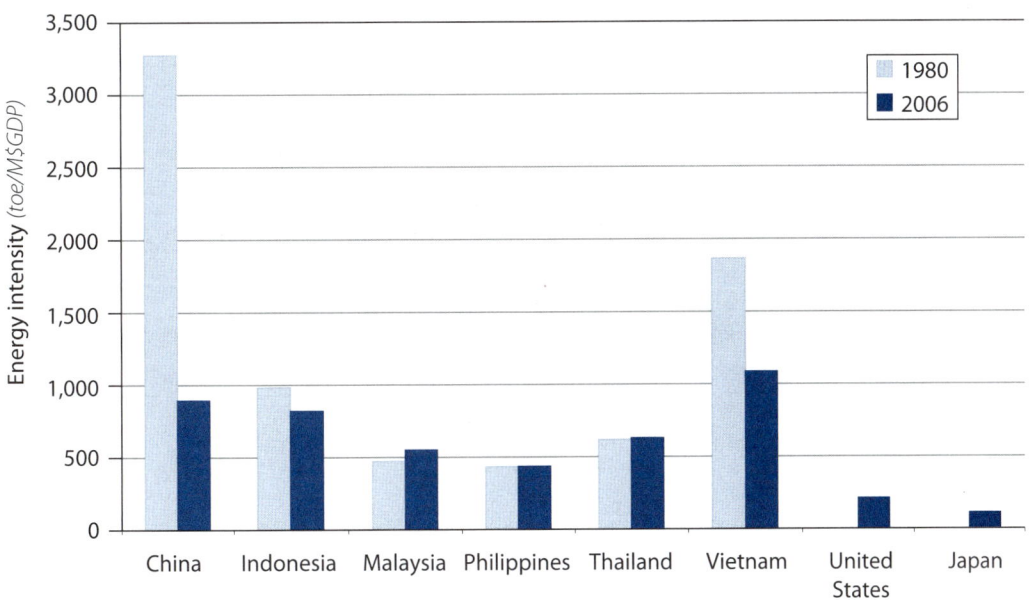

Source: Authors based on IEA data 2008c.

in reducing energy intensity by 70 percent over the last 25 years. Vietnam also has substantially reduced its energy intensity. However, China and Vietnam remain the least energy efficient economies in the region (figure 1.5). Indonesia, Malaysia, the Philippines, and Thailand have made much less progress in improving energy efficiency, and all of them have seen periods of increasing energy intensity. Overall, the region lags developed economies.

East Asia contains some of the largest global energy consumers. However, its per capita energy consumption is still a small fraction of that of developed countries (figure 1.6).

China relies heavily on coal to meet 70 percent of its commercial energy demand. For the EAP5, oil and gas dominate the energy mix, with only 12 percent from coal (figure 1.7). China is the world's largest coal producer and consumer, but coal will not meet all the growth in its energy needs. More than 90 percent of Chinese coal resources are located in inland provinces, but the biggest increase in demand is expected to come from the coastal region. This discrepancy adds to the pressure on internal coal transport and makes import to coastal provinces more competitive. China became a net coal importer in 2007 (IEA 2007). In the longer term, this study suggests that China will become the largest importer of coal in the region.

Figure 1.6 East Asia Has Some of the World's Largest Global Energy Consumers, but Region's Per Capita Consumption Remains Low

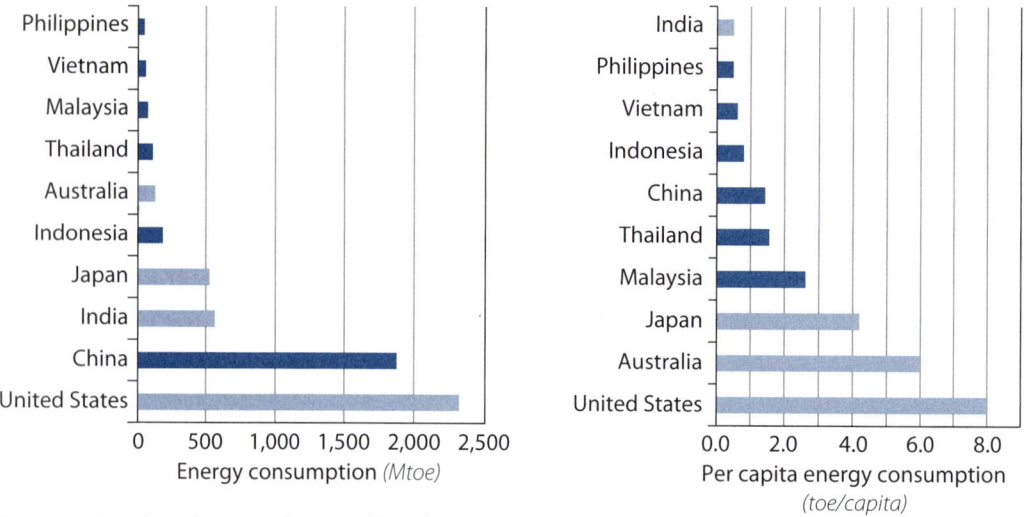

Source: Authors based on data from World Bank 2009c.

Figure 1.7 China Relies Heavily on Coal, Whereas Oil and Gas Dominate EAP5 Energy Mix, 2007 *(%)*

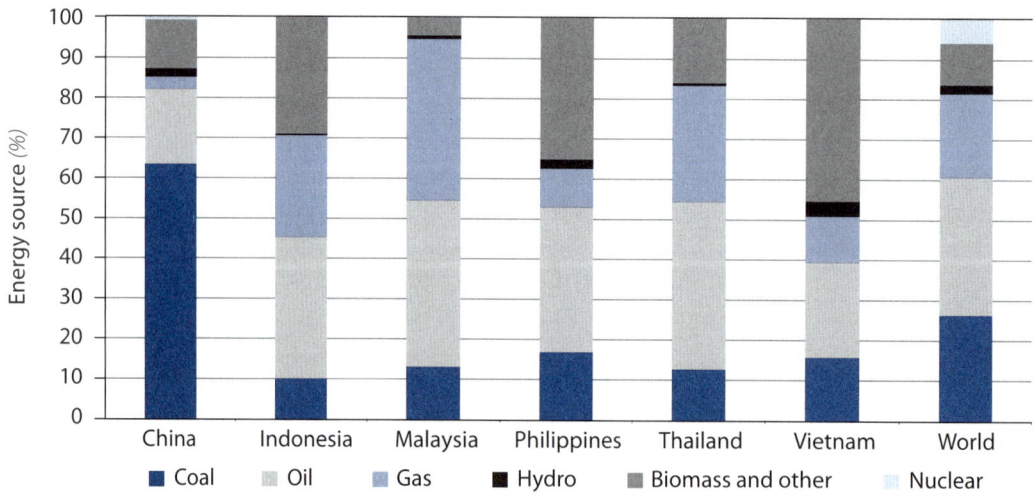

Source: Authors based on data from IEA 2008d.

EAP5 countries have substantially increased their dependence on coal as a reliable source of primary energy over the last three decades. Indonesia has large coal resources. In fact, Australia and Indonesia are the world's largest coal exporters.[15] Hence, coal

---

15. Their exports are worldwide.

exporting countries—Indonesia and Vietnam—have significant incentives to continue increasing the role of coal in their energy mix.
A high dependence on imports to meet the coal demand in Malaysia,
the Philippines, and Thailand is not causing energy security concerns due to a relatively low share of coal in their energy mix and
lesser volatility of coal prices (compared to oil) in the international
market. This analysis shows that EAP5 countries are likely to significantly increase their use of coal unless the environmental costs are
internalized at a sufficient level to reverse this trend.

The region's dependence on oil as an energy source has declined over the past three decades, but with significant variations
among countries. From 1980 to 2006, the share of oil in total primary energy consumption remained flat in China and increased in
Vietnam. However, in Indonesia, Malaysia, and Thailand, the share
was substantially reduced from 80 percent–90 percent in 1980 to
40 percent–50 percent in 2006. EAP5 countries have significantly increased their dependence on natural gas as an energy source
over the past three decades, while in China, the share of natural
gas in total primary energy consumption remained flat at merely
3 percent.

The high energy growth with its current energy mix has led
to twin main energy challenges facing China and EAP5 countries
(figure 1.8):

- Supplying energy in an ***environmentally sustainable*** manner
  that does not adversely impact GDP growth

- Improving the long-term ***security of energy supplies.***

**Figure 1.8 East Asia Faces Twin Energy Challenges
to Sustain Economic Growth: Global and Local
Environment and Energy Security**

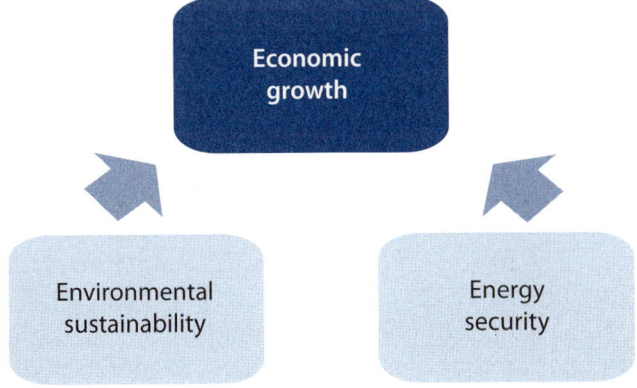

Source: Authors.

Figure 1.9 East Asia Has Many of World's Most Polluted Cities, 2005 *(μg/m³)*

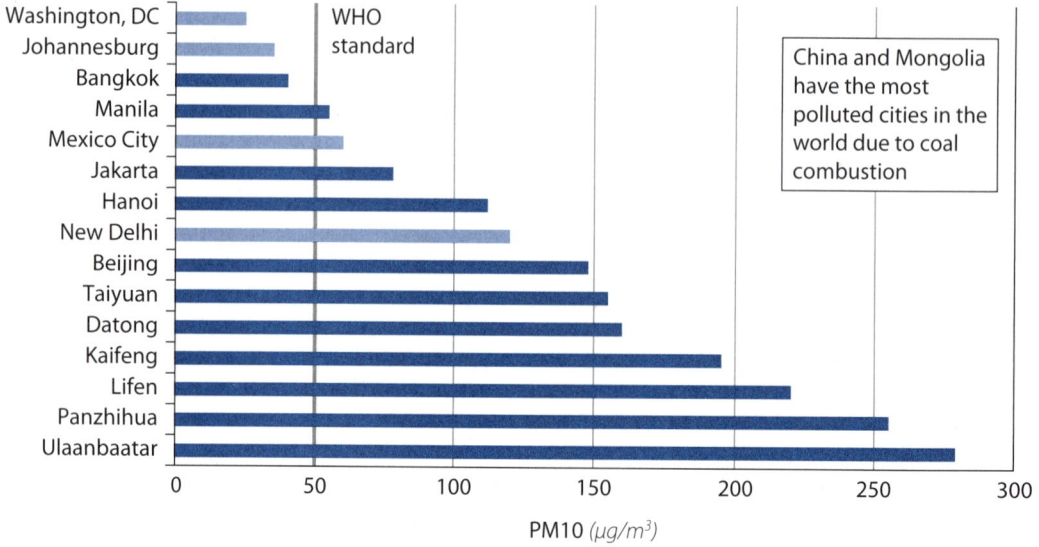

Sources: Authors based on data from National Bureau of Statistics of P.R. China 2006; WHO 2006.

Environmental Sustainability: Global and Local Environment Damages

The East Asia region has paid a heavy environmental price for its remarkable economic growth. Local air pollutants emissions, especially particulates (PM10), sulphur dioxide (SO_2), and nitrogen oxide (NO_x) increased rapidly from coal combustion. As a result, East Asia has many of the world's worst polluted cities, particularly in China and Mongolia due to coal combustion (figure 1.9). China now has 20 of the world's top 30 polluted cities. Urban air pollution from fossil fuel combustion is responsible for 800,000 premature deaths per year globally (Kenworthy and others 2002). Lower-respiratory disease resulting from air pollution is the top burden of disease in the world.

The region's emissions of carbon dioxide have more than tripled over the past 20 years, with China's emissions nearly doubling over the past 6 years (figure 1.10). The region includes top global greenhouse gas (GHG) emitters—China and Indonesia (when emissions from land use changes are counted). Nevertheless, their emissions per capita are lower than those of developed countries (figure 1.11). The developed countries also are responsible for approximately two-thirds of the cumulative energy-related CO_2 now in the atmosphere.[16]

Heavy reliance on coal combustion and vehicle emissions resulting from rapid urbanization are the main drivers of the environmental damage. The scale of these emission increases has raised major

16. WRI 2005.

Figure 1.10 Energy-Related Carbon Dioxide Emissions Grew Rapidly in China and EAP5, 1986–2006 *(MtCO$_2$)*

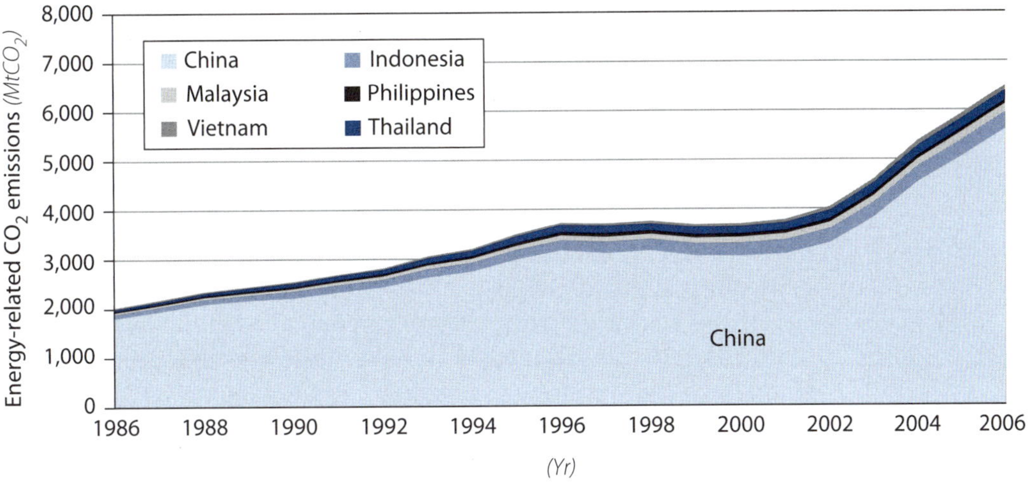

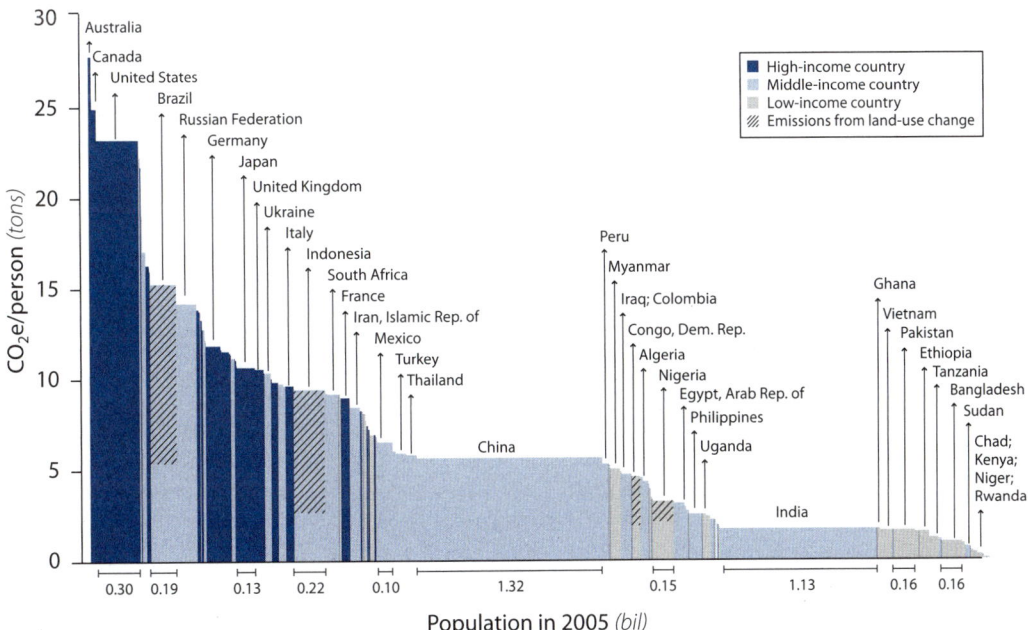

Sources: Authors based on data from IEA 2008c.

concerns in China and elsewhere. The level of air pollution has spurred the installation of pollution abatement equipment on Chinese coal-fired power plants and vehicles. However, this equipment can mitigate the local air pollution problem but not CO_2 emissions.

Figure 1.11 China and EAP5 Countries Have Much Lower CO$_2$ Emissions per Capita Than Industrialized Countries

Source: World Bank 2009a.

Energy Security: Growing Reliance on Oil and Gas Imports and Fuel Price Volatility

Limited oil and gas resources. The East Asia region has a very low share of the world's oil and gas resources—not enough to meet the region's future needs (figure 1.12). Measured in relation to current production (the reserves to production ratio), the region's oil resources are one-third of the world average. The region's gas reserves to production ratio are less than half of the world average (table 1.1). On a per capita basis, the region's oil and gas reserves are only 10 percent of the world average. Given these circumstances, gas production in the region is likely to peak and then fall before 2030.

Gas resources, the least carbon intensive fossil fuel, vary greatly within the region. Indonesia has the largest gas reserves and resources. Indonesia and Malaysia are 2 of the region's largest exporters of gas, meeting 40 percent–50 percent of the regional demand through liquefied natural gas (LNG) exports. Vietnam also has large potential, although yet unproven, gas resources and is self-sufficient in gas supply. China and Thailand are gas importers. The Philippines has limited gas supply and would need to import gas in the future.

Opportunities for gas trade via pipeline within the region are limited but expanding. Singapore imports from Indonesia; and Thailand from Myanmar and, in the future, Malaysia. Thailand also may become an importer from Cambodia once the latter's gas fields are developed. China is becoming a significant importer from Central Asia and also may start large-scale imports from Russia. However,

Figure 1.12 East Asia's Oil and Gas Reserves Are Well below World Averages

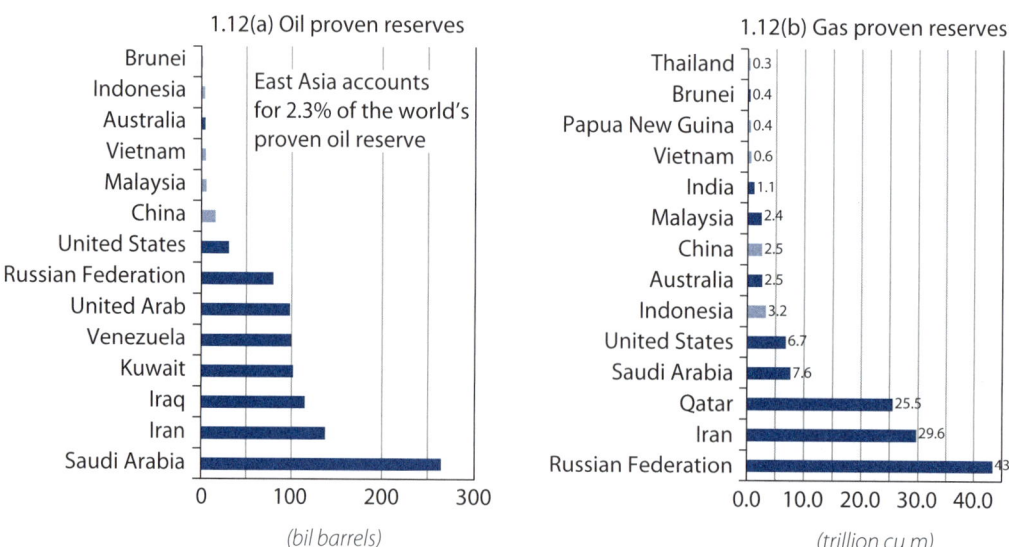

Source: Authors based on data from BP 2009.

Table 1.1 Gas Reserves, Production, and Consumption in East Asia *(bcm)*

Gas	Reserves	Production	Consumption	Imports
China	2455	76	81	5
EAP5	6544	173	117	−56
Indonesia	3184	70	38	−32
Malaysia	2387	63	31	−32
Philippines	112	3	3	0
Thailand	304	29	37	8
Vietnam	557	8	8	0

Source: Authors based on data from BP 2009.

the bulk of gas trade in the entire Asia Pacific region[17] is in the form of LNG, and this is likely to continue.

Gas demand has been growing rapidly in the region at approximately 7 percent per year. Gas accounts for only 7.7 percent of the regional energy mix—low compared to a worldwide average of 24 percent. However, this average hides major regional differences. In Indonesia, Malaysia, and Thailand, gas makes up 20 percent–40 percent of primary energy supply. By comparison, gas comprises only 3 percent of China's energy supply. Given its own limited gas resource, for China to significantly expand its share of gas in total energy, the country inevitably will become a major importer, which will put additional pressure on the region's limited resources.

Fuel switching from coal to gas is particularly important in electricity generation. The current power generation mix in East Asia is 73 percent coal; 9 percent gas; 14 percent hydroelectric; and a little over 1 percent each for oil, nuclear power, and renewables. However, this regional picture is heavily dominated by China, over 80 percent of whose power is generated by coal. In EAP5 countries, coal's share of power generation is only 21 percent, much smaller than the 57 percent share of gas.

Currently, for base load power generation, coal-fired plants are least cost in financial terms. Meanwhile, gas-fired plants may find themselves increasingly uncompetitive, since gas prices are expected to increase in real terms faster than coal prices. Coal-fired plants also offer a better hedge against oil price risk than gas-fired plants, because gas prices correlate strongly with oil prices, but coal prices do not. Therefore, to bring about a greater change in fuel shares, firm policies by governments, such as the incorporation of environmental external costs into coal pricing, will be required.

---

17. Including the major high-income, importing economies of Japan, the Republic of Korea, and Taiwan, China.

Figure 1.13 Reliance on Oil Imports Has Grown in Most East Asian Countries over the Last 25 Years, 1980–2006 *(%)*

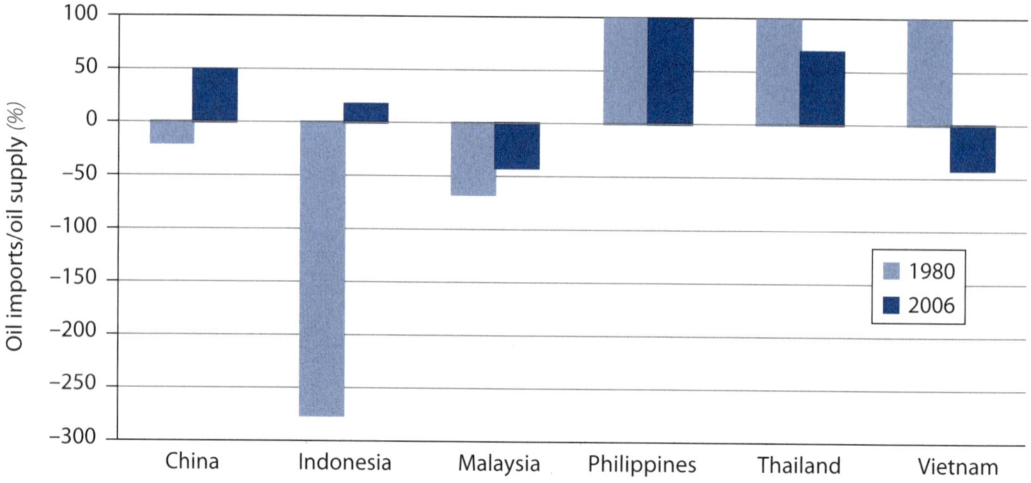

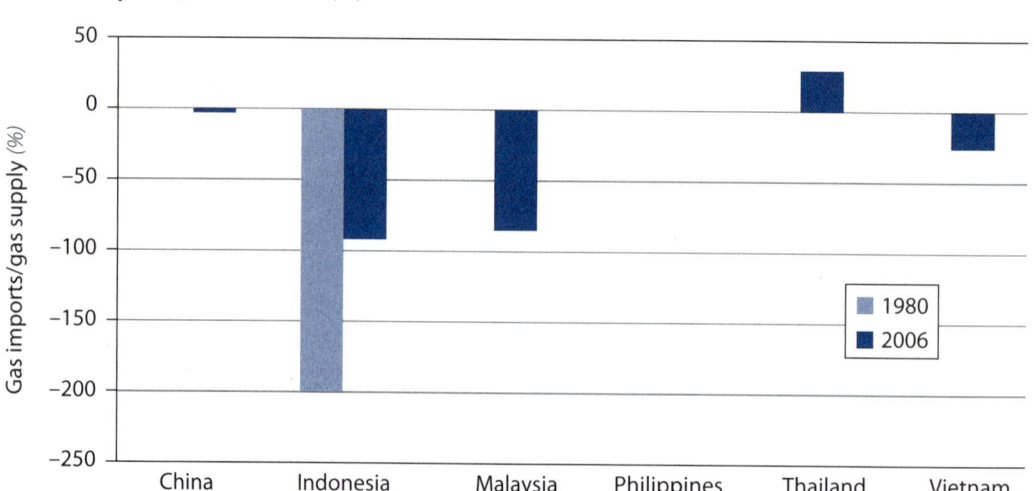

Source: Authors based on data from IEA 2008d.

Oil and gas imports. Over time, China and Indonesia have become more vulnerable to oil. Since 1995 and 2000, respectively, they have become net oil importers (figure 1.13). Malaysia and Vietnam have maintained a stable level of oil self-sufficiency. Over the past two decades, Thailand has decreased its dependence on oil imports, and the role of oil has diminished in the Philippines' energy mix.

Compared to its oil vulnerability, the region is much less vulnerable to gas (figure 1.14). The reason is that gas still represents a relatively small share of the energy mix in China, the Philippines, and

Figure 1.14 East Asia Contains Global Leading Gas Exporters and Is Less Vulnerable to Gas Imports, 1980–2006 *(%)*

Source: Authors based on data from IEA 2008d.

Figure 1.15 Oil and Gas Prices Highly Volatile in Past Decade, 1999–2009 *($/GJ)*

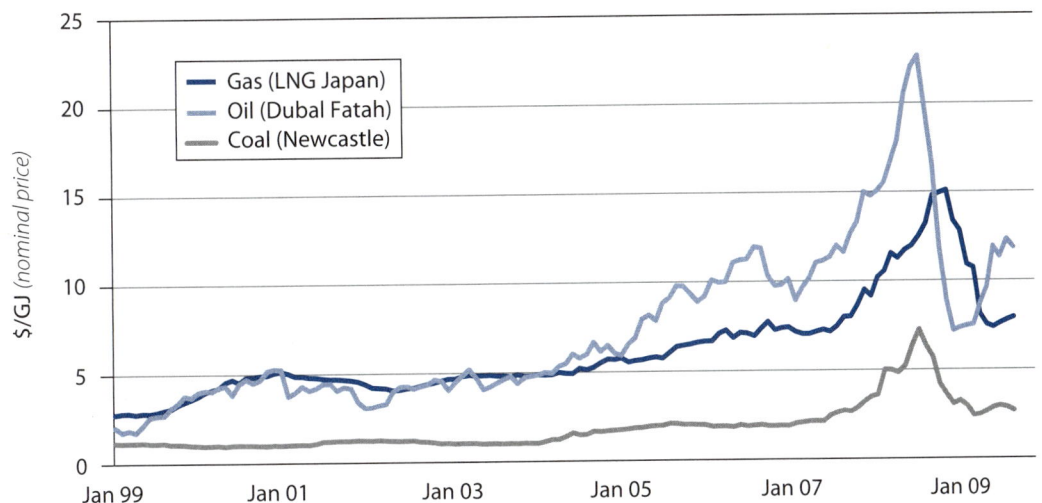

Sources: Authors' calculations based on data from IEA for oil, http://tonto.eia.doe.gov/dnav/pet/pet_pri_wco_k_w.htm; Dubai Fatah crude; GlobalCOAL for coal, http://www.globalcoal.com/default.cfm; NEWC index (Newcastle, Australia); and IEA and Petroleum Association of Japan, http://www.paj.gr.jp/english/index.html, for Japan LNG.

Vietnam. Indonesia and Malaysia are the world's leading gas exporters. In these two countries, gas accounts for a significant share of the total primary energy supply. Only Thailand has similar patterns of oil and gas vulnerability. These figures indicate significant room for the possible expansion of gas use in the region in response to energy security and environmental concerns, but under the constraints of locally available resources and long-term marginal costs of gas production and transport.

Fuel price volatility. Its limited oil and gas resources imply that the region will rely increasingly on oil and gas imports to meet its growing energy needs. The East Asian countries are already significant energy traders and include some of the world's largest exporters and importers. Their energy trading, in turn, makes these countries more vulnerable to international fuel price volatility, which many of them find politically difficult to pass on to consumers (figure 1.15).

2

Energy Scenarios

Key messages: The study underlying this report examined two energy scenarios up to 2030: (1) a reference scenario, which features a continuation of current government policies (REF scenario); and (2) an alternative scenario of sustainable energy development (SED scenario), which aims to put the energy sectors on a sustainable path. Two separate studies projected the growth of energy demand in the transport and household sectors. The transport study examined the potential to reduce transport fuel consumption through fuel economy standards, public transport, urban planning, and pricing policies. The household study explored the potential to reduce residential electricity consumption through appliance efficiency. The findings of these studies have been incorporated into the Policy Assessment Model, developed for this study to analyze alternate scenarios. Finally, this study also conducted sensitivity analyses to evaluate the impacts of What if energy efficiency improvement were less than what is assumed in the SED scenario? and What if new technologies could reach sizable scale prior to what is assumed in the SED scenario?

2.1 Overall approach

To address the twin challenges of environmental sustainability and energy security, the study examined policy impacts of future energy scenarios up to 2030. These scenarios are not forecasts or energy plans. Their purpose is to enable policymakers to obtain a better understanding of the quantitative impacts of policy options. This study focuses on the comparison of two main scenarios:

- *Reference (REF) scenario.* The REF scenario assumes a continuation of current government policies, making adjustments where necessary to project them forward to 2030 and to implement what are considered to be more realistic options in light of recent economic conditions. The REF scenario presumes that most governments' plans and targets will be realized, without judging the likelihood of their being achieved. This assumption may underestimate the emissions under the REF scenario, since some governments' clean energy targets may not be achieved.

- **Sustainable Environment Development (SED) scenario.** A set of policies aimed at delivering more environmentally sustainable outcomes. SED represents the maximum technically feasible changes that could most likely be achieved. The SED scenario represents a low-carbon growth path with least abatement cost under the constraints of maximum technically feasible potential of energy efficiency and low-carbon technologies. SED considers primarily proven technologies and excludes carbon capture and storage (CCS). While CCS is critically important in the long term (beyond 2030), particularly for China, it is expected to be commercially available only after 2025. Furthermore, in China, the REF scenario is based on historical energy efficiency efforts and government's targets of renewable energy and nuclear power announced in 2005. The SED scenario extrapolates government's most recent EE, RE, and nuclear power targets to 2030.

In reality, it is challenging to achieve the maximum technical potential of energy efficiency measures under the SED scenario, and new technology development is uncertain in the future. To this end, this study conducted sensitivity analyses by developing three additional scenarios to evaluate the impacts of (1) What if energy efficiency improvement could be less than what is assumed in the SED scenario? and (2) What if new technologies could reach sizable scale prior to what is assumed in the SED scenario? The three additional scenarios are:

- **SED and aggressive advanced technologies.** The third scenario is a combination of SED and an upper boundary of CCS and solar energy. To gain insight into the role and timing of advanced technologies, this scenario assumes deployment of CCS and solar energy at a maximum level. This assumption is based on CCS technology becoming commercially available on a large scale by 2020 and the costs of solar energy being significantly reduced.[18]

- **Lower energy efficiency**. To examine the underlying uncertainty and role of energy efficiency, the fourth scenario assumes energy efficiency improvement would be halfway between the assumptions in the REF and the SED scenarios, while keeping other assumptions the same as those in the SED scenario.

---

18. It is assumed that CCS would account for 25% and 10% of the coal-fired power capacity in China and each of the EAP5 countries, respectively, by 2030. Solar PV capacity would reach 150 GW in China and 2 GW in each of the EAP5 countries by 2030.

- *Lower energy efficiency and aggressive advanced technologies.* To assess the trade-off in roles and timing of energy efficiency vs. advanced technologies, the fifth scenario assumes a combination of a lower energy efficiency improvement level (halfway between the assumptions in the REF and the SED scenario) and deployment of CCS and solar energy at a maximum level.

The assumptions underlying future policy options in the power sector are based on (1) evaluation of the scope for increased gas use in the region; (2) estimate of the potential for energy efficiency improvements; and (3) development of power-generating technologies focusing on the costs and speed of deployment of renewable energy, nuclear power, and clean coal technologies (advanced high-efficiency coal technologies and carbon capture and storage).

Two separate studies were conducted for transport energy and household electricity demand. The transport study examined the growth of energy demand and the potential to reduce future transport fuel consumption through fuel economy standards, implementation of city transport master plans of public transport and urban planning, and road pricing and fuel tax policies. The study looked at major urban areas within the large emerging economies of the East Asia region, and extrapolated the results to a country-wide basis for the scenario analysis. The household energy study projected the growth of residential electricity demand and appliance ownership in the Philippines, Thailand, and Vietnam. The study examined the potential to reduce household electricity consumption through improving the efficiency of appliances.

The findings of these studies were incorporated in a policy assessment model (PAM) developed specifically for this study to analyze the alternate scenarios (figure 2.1). The PAM projects energy demand, supply, costs, and emissions for each EAP country by fuel and by sector, including the scope for improved energy efficiency on final energy demand in the industry, transport, and residential/commercial sectors.

Since the power sector is the largest single source of CO_2 emissions, all the leading climate and energy models project decarbonizing the power sector first.[19] Hence, the PAM power sector analysis is a key part of the methodology because a major share of the investment substitutions between REF and SED scenarios occur in the power sector. The power sector analysis is performed in two parts: (1) the development of the power generation scenarios to 2030 to determine the investment in new capacity and (2) the simulation of the least-cost dispatch of the power plants to meet demand.

---

19. IPCC 2007; IEA 2008b; Calvin and others 2009; IIASA 2009.

Figure 2.1 Overview of the Policy Assessment Model

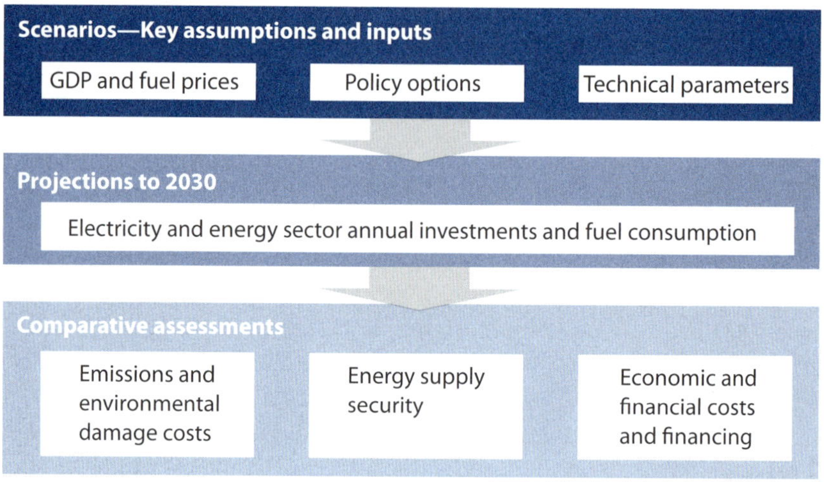

Scenarios—Key assumptions and inputs		
GDP and fuel prices	Policy options	Technical parameters

Projections to 2030
Electricity and energy sector annual investments and fuel consumption

Comparative assessments		
Emissions and environmental damage costs	Energy supply security	Economic and financial costs and financing

Source: Authors.

The power generation scenarios for each country take account of the supply possibilities, the existing system and government policies, the shape of the load curve, demand expansion, technical constraints, and the comparative economics of each technology. The supply possibilities were analyzed through projections of each country's energy resources, the technical-economic potential of each of the new renewable and other technologies, and the Asia-Pacific regional market context for import options. Screening curves were used as a first step to identify the least-cost set of power plant investments to meet the shape of the load curve, that is, base load, mid-merit, and peaking plant. Finally, for the REF and SED scenarios, the relevant policy interventions for the long-term fuel mix were used to guide the final determination of annual investment in each technology.

The power generation from each plant and technology type was calculated by carrying out an annual least-cost dispatch using a load curve for the year. Technical and policy constraints were reflected in the minimum and/or maximum levels of annual dispatch for each plant type. This calculation produces an annual projection of fuel consumption, fuel costs, and other operating costs. These are added to the investment costs from the power generation scenarios and the capital costs of the existing capacity to give the total financial costs of power generation and unit electricity costs. The dispatch model also projects the local and global emissions of CO_2, SO_2, NO_x, and total suspended particulates (TSP) which are used, together with the unit damage costs, to estimate the total economic cost of power generation.

2.2 Scenario Assumptions

The key assumptions underlying the REF and SED scenarios are compared in table 2.1.

Table 2.1 Key Assumptions of REF and SED Scenarios

	REF scenario	SED scenario
GDP growth	Annual growth 2007–30: China 6.8%, Indonesia 5.0%, Malaysia 5.0%, Philippines 4.2%, Thailand 4.4%, Vietnam 6.0% IMF 2006 and APERC 2006	
Fuel prices[1]	Oil rising to $85/bbl ($13.4/GJ) by 2030 Gas rising to $13.2/mmBtu ($12.5/GJ) by 2030 (linked to oil price) Coal rising to $85/t ($3.1/GJ) by 2030	
Income and price elasticity	Derived from review of recent studies	
Domestic production	For China, as *IEA World Energy Outlook 2008*. For other countries, as most recent country forecasts or, where not available, as APERC 2006.	
Power generation capacity	The same as the most recently published power development plan, which was adjusted for excess reserve margins (in China, this capacity is based on the government's targets of renewable energy and nuclear power announced in 2005)	Maximum technical feasibility of renewable energy based on resource potentials in each country, and major expansion of nuclear power in China (in China, this capacity is based on the most current government targets for renewable energy and nuclear power)
Nontransport energy efficiency[2]	Annual 1% autonomous improvement (1.5% for industrial use in China)	Energy efficiency increased by an additional 1% improvement annually
Transport energy efficiency	50% of savings under "Do Maximum" strategy identified in transport study (appendix 1)	Extrapolated from outcome under "Do Maximum" strategy identified in transport study (appendix 1)

Source: Authors.
Notes:
1. Assuming international border prices for fuels may overestimate the energy efficiency fuel savings, because in reality, fuel subsidies exist in several EAP countries. In addition, these fuel price projections may be lower than other international estimates. However, if higher fuel prices were used, demand would fall in both REF and SED scenarios; therefore, this assumption will not change the comparison of the two scenario results.
2. Autonomous energy efficiency improvements in each end-use sector reduce the energy required to produce the same amount of output through technological changes, assuming other variables, such as energy prices, are fixed.

Table 2.2 Key Assumptions of Sensitivity Analyses

	SED and aggressive advanced technologies	Lower energy efficiency	Lower energy efficiency and aggressive advanced technologies
Nontransport energy efficiency	Annual 2% autonomous improvement	Annual 1.5% autonomous improvement	Annual 1.5% autonomous improvement
CCS deployment	Upper boundary of CCS deployment after 2020	No significant deployment by 2030	Upper boundary of CCS deployment after 2020
Solar energy deployment	Upper boundary of deployment after 2020	Limited deployment by 2030	Upper boundary of deployment after 2020
Power generation capacity	Maximum technical feasibility of renewable energy based on resource potentials in each country and major expansion of nuclear power in China. (In China, this assumption is based on government's most recent targets for renewable energy and nuclear power.)		

Source: Authors.

Table 2.2 compares the key assumptions underlying the three additional scenarios of sensitivity analyses.

2.3 Scenario indicators

Environmental Sustainability

Our analysis considers both the global and local[20] environmental impacts of energy use. Global environmental impacts are measured as the total annual emissions in CO_2 from fossil fuel combustion. Emission intensity (emissions per unit of GDP) and emissions per capita also are reported to enable comparisons of different countries. The cost of carbon emissions is assumed to be $20/tCO_2$.

Economic damage costs associated with three air pollutants—TSP, SO_2, and NO_x—are used as a single combined measure of local environmental impacts, which are limited to those associated with power generation—the most significant source of these impacts. This study adopted the benefit transfer method to estimate local environmental damage costs. This method was used in the economic analyses of Bank investment projects in East Asia countries. This method transfers the unit damage cost of dollar per person per ton emission derived from the well-known environmental impact study of power plants in New York State[21] to the unit damage cost for China and the EAP5 countries. The transfer is based on the difference in purchasing power parity (PPP) per capita. The unit cost of per ton emission is then calculated by multiplying the unit damage cost by the number of persons in the

---

20. "Local" is used to distinguish this category of environmental damage from the global impacts of carbon emissions. Some local pollutants—notably sulfur dioxide (SO_2)—can have impacts over a wide geographic area.
21. The New York study estimated the value of the statistical life of both mortality and morbidity from air pollution.

affected population. In China, for example, the national average local environmental damage costs for TSP, SO_2, and NO_x, weighted by PPP per capita in each province, are estimated at \$1465/ton, \$335/ton, and \$373/ton, respectively. It is assumed that these figures will increase proportionally to GDP per capita in the future. It is important to note that there is considerable uncertainty over these damage costs.

Energy Security

The main indicators to assess security of supply are (1) fuel diversity; and (2) import dependency and absolute amount of imports, particularly oil and gas imports, to reflect the risk of energy supply disruptions and price volatility.

Higher levels of fuel diversity generally will be associated with higher supply security. In other words, a physical or financial disruption in supplies of one fuel has a limited impact on the economy as a whole. To capture this correlation, the selected measure of fuel diversity is an Inverse Simpson (IS) Index.[22] The calculated index is based on five primary fuel groups: oil, gas, coal, nuclear, and renewables. With 5 fuels, the IS Index has a maximum score of 5 (equal share for each fuel group) and a minimum score of 1 (100 percent share for 1 fuel group).

High levels of import dependency usually, but not always, are associated with lower supply security.[23] Import dependency is measured as the share of imports (in physical units) in total primary energy supply. Since oil and gas prices are most volatile, this study focuses on oil and gas dependency.

Costs and Financing

This study estimated three indicators of the average financial costs: (1) average cost of primary energy supply, (2) average cost of electricity generated, and (3) total cost as a share of national income (GDP). In addition, for developing countries, financing the upfront investment costs of energy efficiency and low-carbon technologies is a major challenge. Hence, this study also estimated (1) additional investment costs required for energy efficiency, renewable energy, and nuclear power for the SED scenario compared with the REF scenario; and (2) the amount of concessional financing required for energy efficiency and renewable energy.

---

22. This index is calculated as "one over the sum of the squares of the percentage share of each fuel." This is essentially the same as the Hirschmann-Herfindahl Index, a commonly accepted measure of market concentration.

23. Where domestic production is vulnerable to disruption due to industrial action or transport bottlenecks, imports may be a more reliable supply source.

The total financial costs of the SED scenario are calculated as annualized capital investment costs of new electricity generation capacity, total fuel supply costs, and the costs of incremental energy efficiency investments in the industrial, transport, and residential/commercial sectors. The average cost of primary energy supply is calculated as the average fuel cost, using border prices for internationally traded fuels, which ignores the costs of transport and distribution within a country.[24] In both REF and SED scenarios, the average cost of electricity generated is the sum of the average annualized capital investment costs of new generation capacity, operations and maintenance costs, and average fuel costs. Transmission and distribution investment costs are excluded.[25]

Technology breakthroughs and economies of scale can lead to rapid cost reductions of new technologies. This study adopts a dynamic analysis that factors in the potential of cost reduction for new technologies such as many renewable energy technologies. This analysis is based on the assumptions of learning rate: the cost reductions associated with doubling capacity.

This study calculated the additional investment costs in energy efficiency for the SED scenario compared to the REF scenario. Given the wide range of energy efficiency projects that could be undertaken, the costs of energy efficiency investments are difficult to determine. According to IEA, an additional $1 invested in demand-side electricity avoids more than $2 in investment on the supply side (including generation, transmission, and distribution). This ratio varies by geographic region. In the OECD countries, the ratio is $1 invested to $1.6 avoided. In developing countries, the ratio is larger, at $1 invested to more than $3 in supply costs saved (IEA 2006). This study used this relationship to derive an implicit aggregate energy efficiency cost curve for each country. In the initial years, energy efficiency opportunities would achieve a return of a payback period of 3.3 years. As time progresses, more of the easier investments will have been made, leading to a rise in the payback period to 5 years. The avoided fuel cost of saved electricity was used to estimate the energy efficiency investment cost as this would be the variable producing the return on supply-side investment in

24. Border prices are world market benchmark prices plus costs of transport to the country concerned. Countries may choose to price fuels below world market levels. Such decisions (fuel subsidies) are not taken into account in the comparisons of scenarios and country performance due to their distortionary effect.

25. The model's exclusion of investments in power transmission and distribution underestimates the difference between REF and SED investment costs and makes SED investment estimates quite conservative. For this reason, the SED scenario has a lower electricity demand (due to a higher energy efficiency level) and would require fewer investments in power distribution than the REF scenario. Typically, investments in power distribution are of the same magnitude as investments in power generation, and the difference between investments in power distribution between SED and REF is not small.

electricity generation. This is considered to represent a reasonable range of costs of energy efficiency investments that would be undertaken in practice. The investments in the transport sector only include capital costs of efficient vehicles, not those of public transport infrastructure. Therefore, it underestimates the investment needs in the transport sector.

Although many energy efficiency measures have short payback periods, they often face financing barriers. Individual consumers usually demand very short payback time and are unwilling to pay higher upfront costs for energy-efficient products. Financial institutions usually are not familiar with or interested in energy efficiency financing, because of the small size of the deal, high transaction costs, and high perceived risks. Therefore, concessional financing is required to cover the incremental risks, financial incentives, and transaction costs, in addition to grants for capacity building of local stakeholders. Without these interventions, the technical potential of energy efficiency would be difficult to achieve in reality.

This study estimated the order of magnitude of this required concessional financing as a percentage of the total additional investment costs in energy efficiency between the REF and SED scenarios. The authors estimated the investment costs to cover the risk premium between the payback period for energy efficiency investments and consumers' willingness to pay (the payback time at which consumers are willing to invest) in each end-use sector. This calculation is used as an approximate indicator for risk mitigation and financial incentives required.

The payback period at which consumers are willing to invest varies by sector. The residential sector usually has the lowest paybacks of 1.5–3.0 years, and households often require financial incentives for energy efficiency. The industrial sector, the largest source of energy efficiency investment, usually is willing to invest in energy efficiency projects with 2–4 years of payback, and risk mitigation usually plays a key role in promoting energy efficiency. The transport sector usually has the highest payback period of 3.0–4.5 years. In addition, the study examined the WBG portfolio of energy efficiency financing and risk guarantee as empirical evidence to estimate the technical assistance required to enable local financial institutions to mainstream energy efficiency lending.

Renewable energy technologies are capital intensive but with zero or low fuel costs, compared to coal-fired power plants, which are the assumed baseline in this study. On the other hand, many renewable energy technologies (except geothermal) have lower capacity factors than coal-fired power plants.[26]

---

26. Capacity factor is defined as the ratio of the actual output of a power plant over a period to a plant's output if it had operated at full capacity during the entire period.

Figure 2.2 Environmental Sustainability, Energy Security, and Costs of REF Scenario, 2010

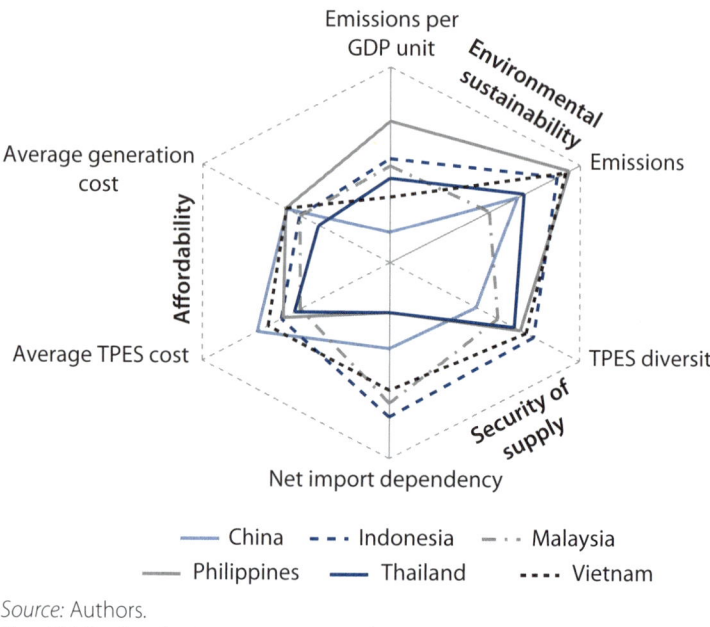

Source: Authors.
Note: TPES = "total primary energy supply."

This study estimated the order of magnitude of the required concessional financing as a percentage of the total additional investment costs in renewable energy between the REF and SED scenarios. First, the study calculated the investment costs of coal-fired power plants to deliver the same power output of electricity generation as renewable energy as the baseline financing cost. Next, the fuel savings of avoided fossil fuel costs from renewable energy were estimated. The required concessional financing is the result of the total additional investment costs in renewable energy moving from REF to SED scenarios minus baseline financing and fuel savings.

An overall comparative assessment of the indicator results is presented visually using radar diagrams. The relative performance of a scenario against an indicator is represented by the radial distance from the center of the diagram. Using this approach, the position in 2010 for China and EAP5 countries under the REF scenario can be seen in figure 2.2.

3

Baseline: Unacceptable Environmental Damages and Growing Energy Insecurity

Key messages: Continuing current policy is not sustainable. Sustaining economic growth without compromising the environment is expected to be the greatest energy challenge facing East Asia over the next two decades. From 2010 until 2030, energy demand will double to meet the high growth rates. China will continue to rely heavily on coal, whereas the share of coal will decline slightly. The EAP5 countries plan to significantly expand the role of coal. As a result, emissions of CO_2 and local air pollutants also will double over the next 20 years. Climate change impacts and local air pollution will take a heavy toll on growth and development. The major emerging economies in the region likely will face carbon constraints by 2030. The region's reliance on oil and gas imports also will grow, increasing energy security risks of price volatility and energy supply disruptions. Therefore, economic growth cannot be sustained.

3.1 Energy demand will double, and coal will continue to dominate

Based on the annual growth rates shown in table 2.1, regional GDP is expected to increase 4-fold by 2030. Under the REF scenario, energy use will double from 2010 to 2030 (figure 3.1). In comparison, under the IEA's projection, world energy demand is expected to grow by 45 percent by 2030 (IEA 2008a).

The fact that the energy growth rate is half of the GDP growth rate indicates that, under current policies, East Asian countries already are taking significant steps to improve energy efficiency and reduce their economies' energy intensities. The REF scenario

47

Figure 3.1 Under REF Scenario, East Asia's Energy Demand Will Double, 2010–30 *(Mtoe)*

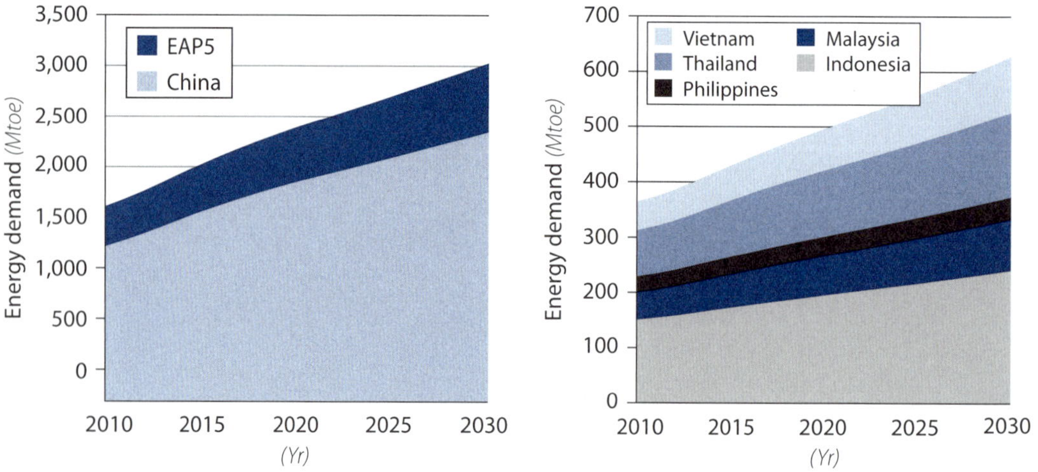

Source: Authors' calculations.

assumes that China's energy intensity would drop by 3.4 percent per year from 2010 to 2030, similar to the achievements over the past decade.[27] The Philippines and Vietnam are expected to reduce their energy intensity by 1.9 percent and 1.6 percent per year, respectively, from 2010 to 2030, less than what they achieved over the past decade (1995–2005). Contrarily, Indonesia, Malaysia, and Thailand need to reverse their trends of rising energy intensity over the past decade to, instead, reduce energy intensity by 2.5 percent, 1.4 percent, and 0.9 percent per year, respectively, from 2010 to 2030.

Under the current government policies, China will continue to rely heavily on coal, while the share of coal will decline slightly from the current 64 percent to 60 percent by 2030. In contrast, as coal is abundant in the region and provides low-cost and secure energy supplies, EAP5 countries plan to significantly expand the share of coal from the current 12 percent to 21 percent by 2030. In China, oil and gas demand is expected to increase faster than coal demand, whereas in EAP5 countries, growth of coal will outpace oil and gas. As a result, coal will continue to dominate the primary energy supply in the region.

By 2030, growing demand for electricity will triple total installed capacity from the current 811 GW to over 2,300 GW. Renewable energy (hydro, biomass, geothermal, wind, and solar) will grow by 50 percent from 2010 to 2030, largely from hydropower. This increase in renewable energy represents the smallest projected increase of all the fuels.

---

27. Energy intensity declined by 6.5% per year from 1995 to 2000. It then increased from 2002–05. Therefore, over that decade, China reduced energy intensity by an annual average of 3.4%.

Nuclear power capacity will experience a major boost via a 12-fold increase from the current very low levels over the next 20 years, due largely to the Chinese government's aggressive plans. The governments of Malaysia, Thailand, and Vietnam also include nuclear power in their medium- and long-term power development plans after 2020.

3.2 CO_2 emissions and local air pollutants also will double

Over the next two decades, sustaining economic growth without compromising the environment is expected to be the greatest energy challenge facing East Asia. The EAP region is among the most vulnerable in the world to climate change threats (box 3.1). By 2030, the major emerging economies in the region likely will face carbon constraints. Over the next two decades, emissions of both CO_2 and local air pollutants will double. China will account for more than 85 percent of this increase, with the balance coming from EAP5 countries (figure 3.2). Among EAP5 countries, Indonesia will account for 40 percent of the emissions, followed by Thailand, Malaysia, Vietnam, and the Philippines in this order.

China and EAP5 countries will reduce their carbon intensity from 2010 to 2030. The most significant cut will be 54 percent by China, followed by 35 percent by Indonesia, 28 percent by Malaysia, 25 percent by the Philippines, 22 percent by Thailand, and 11 percent by Vietnam. China is determined to improve both supply and demand-side efficiency and switch from coal to gas, nuclear power, and renewable energy technologies, whereas the EAP5 countries plan to expand the role of coal.

Figure 3.2 CO_2 Emissions Will Double for All East Asian Countries by 2030 *(MtCO₂)*

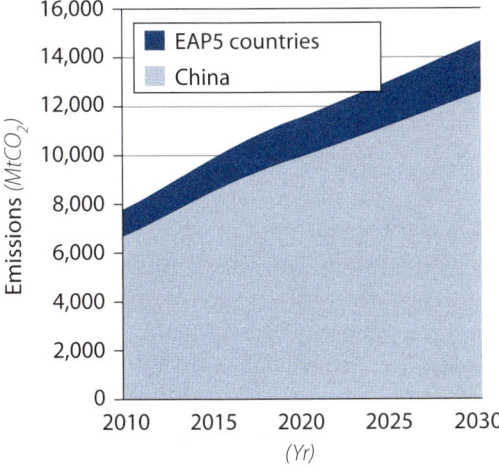
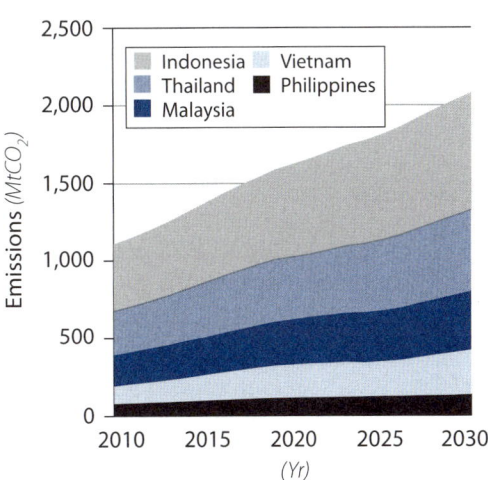

Source: Authors' calculations.

Box 3.1 EAP Is among World's Most Vulnerable Regions to Climate Change Threats

The *huge numbers of people living along the coast and on low-lying islands* are particularly vulnerable to climate change. These people comprise over 130 million in China; roughly 40 million, or more than half of the entire population, in Vietnam; and most Pacific Island Countries (World Bank 2009a). Consequently, the IPCC listed the Mekong Delta as 1 of the top 3 hot spots in the world for potential migration (IPCC 2007).

A second vulnerability is *crop yields*, which are projected to decline in many Asian countries due to rising temperatures and extreme weather events. In the Mekong River basin, the rainy season will see more intense precipitation, while the dry season will lengthen by two months.

Third, the region's economies are highly dependent on *marine resources*. The value of well-managed coral reefs is $13 billion in Southeast Asia alone. These reefs already are stressed by coastal development, overfishing, agricultural pesticide and nutrient runoff, and industrial pollution.

The 2006 Stern Review estimated that if the world community did not act quickly, the overall costs and risks of climate change could equal at least 5 percent of global GDP each year (Stern 2006), and might even be higher for vulnerable PIC countries and vast areas in other EAP countries.

Source: World Bank 2009a.

Emissions of local air pollutants of particulates, SO_2, and NOx also will substantially increase (figure 3.3). Under the REF scenario, local environmental damage costs are estimated at $127 billion in 2030, of which 98 percent will be in China due to its heavy reliance on coal. REF assumes that all East Asian countries will comply with the requirements to install and operate air pollution abatement equipment in all existing and newly built plants. In reality, some countries have not always fully operated the abatement equipment to achieve maximum emission reduction.

3.3 Reliance on oil and gas imports will grow

As concerns over security of supply grew and reached high levels during the peak fuel prices of 2008, all countries examined options for increasing their fuel diversity to strengthen the security of their supplies. Under the REF scenario, the fuel diversity in the EAP5 countries, measured by the IS Index, improves over time. However, the diversity increase will be achieved by expanding coal consumption. For example, coal use in Indonesia is projected to quadruple from 2010 to 2030. Fuel diversity in China will remain low.

Figure 3.3 Local Air Pollution in EAP Region Also Will Double by 2030 *(Kton)*

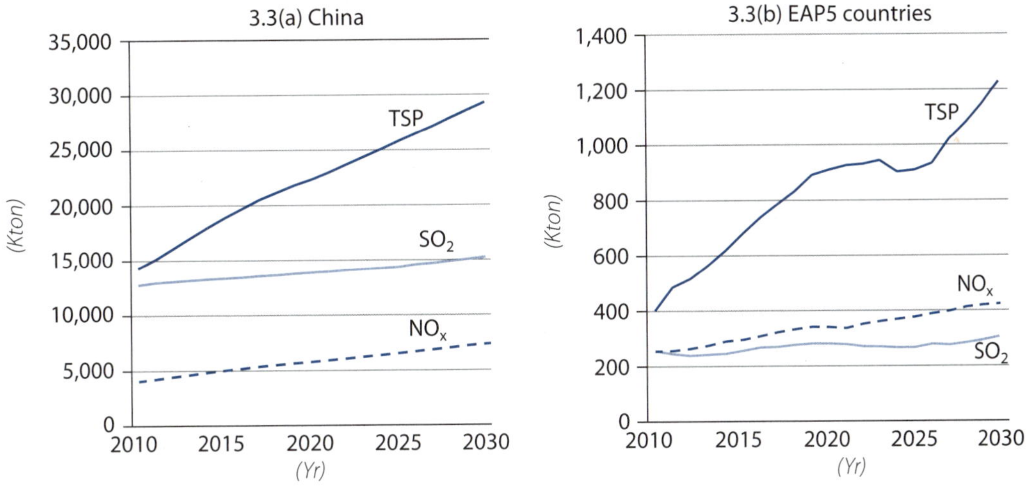

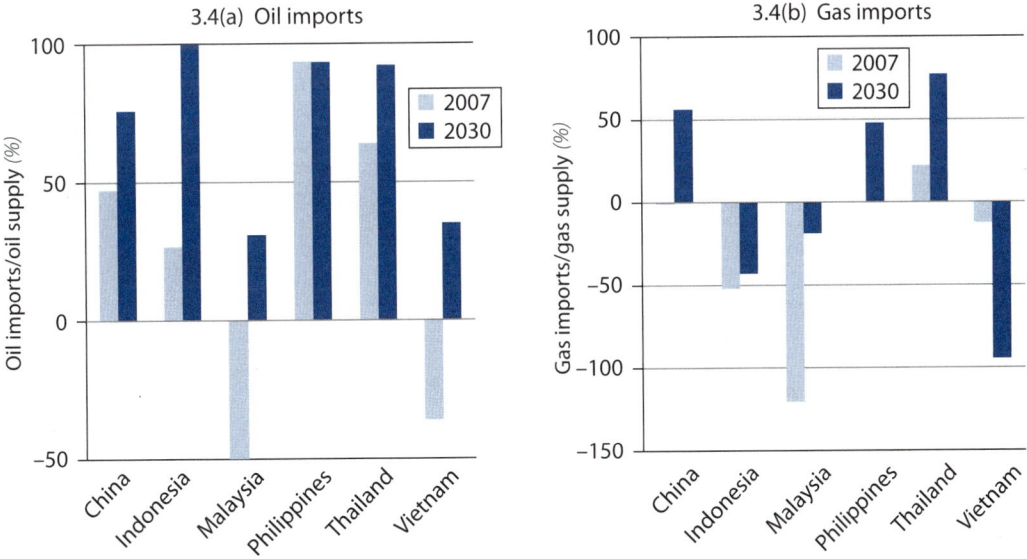

Source: Authors' calculations.

Imports of oil and gas will grow across the region throughout the two decades (figure 3.4). The increases will heighten concerns over energy security and bring increased risks of price volatility and exposure to disruptions in energy supplies. Under the REF scenario, China is expected to import 75 percent of its oil and 50 percent of its gas demand by 2030, thus becoming the world's largest oil importer. Malaysia and Vietnam are projected to switch from being net energy

Figure 3.4 All East Asian Countries Will Increase Reliance on Oil and Gas Imports, 2007–30 *(%)*

Source: Authors' calculations.

exporters to net importers. Thailand and the Philippines are expected to import 60 percent–70 percent of their energy needs.

3.4 Power expansion requires large investments and produces high environmental costs

Under the REF scenario the total capital investment costs (undiscounted) in power generation from 2010 to 2030 will be $2 trillion, or an average of $100 billion per year. Of the annual average, $60 billion will be for thermal power, $30 billion for renewable energy, and $10 billion for nuclear power. Almost 90 percent of this total investment will be in China. The financial costs of REF, including both investment and fuel costs, will be large but will decrease as a percentage of GDP over time.

A comparison between the financial costs of energy supply (investment costs plus fuel costs) and economic costs (including the external environmental costs) highlights the growing importance of the latter (figure 3.5). The economic costs of local environmental damages, although small relative to the financial costs, increase over time as local damage values grow with per capita income. Emission costs are proportionately higher for China, for which coal comprises a far higher proportion of total primary energy supply (currently, 64 percent compared to 12 percent in EAP5).

In sum, the radar diagrams (figure 3.6) show the changes in indicators of environmental sustainability, energy security, and costs under the REF scenario in each major East Asian economy in 2030, compared to the 2007 levels. Higher scores (farther distance from the center) indicate better performance (lower emissions, higher fuel diversity, lower import dependency, and lower cost).

Figure 3.5 Impacts of Incorporating External Environmental Costs in 2030 *(US$bil)*

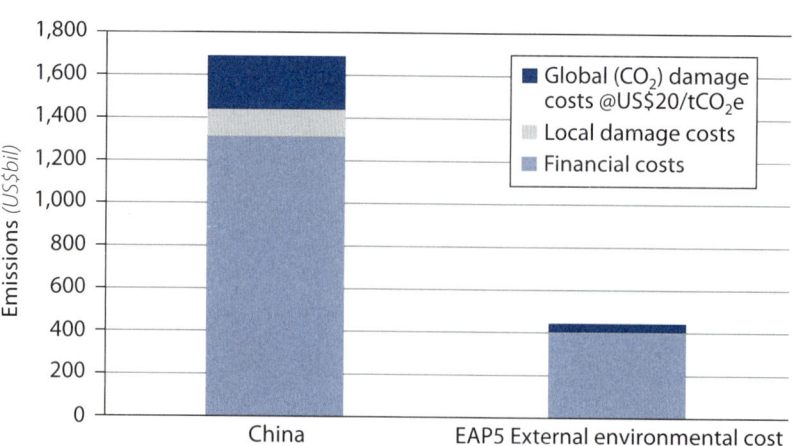

Source: Authors' calculations.

Figure 3.6 Environmental Sustainability, Energy Security, and Costs of REF Scenario, 2030

China

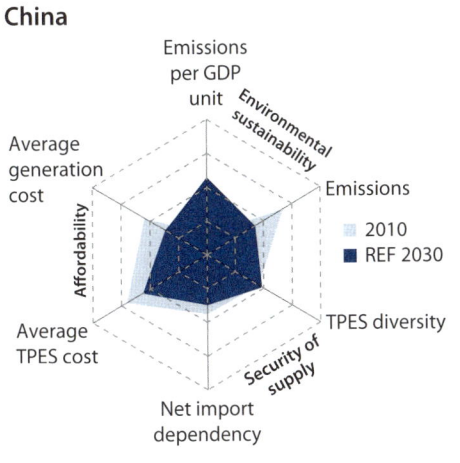

Indonesia

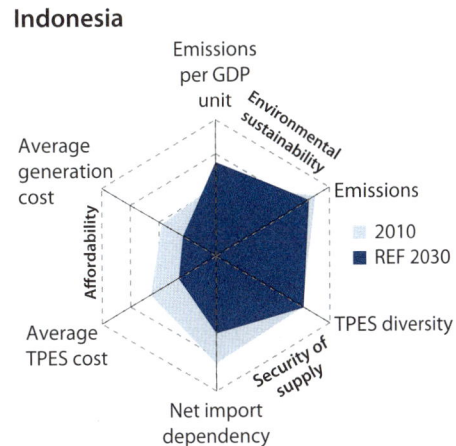

Malaysia

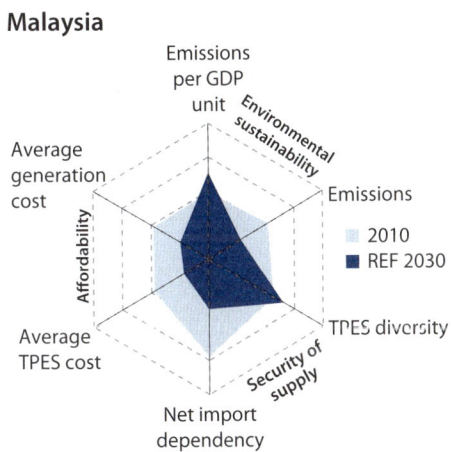

Philippines

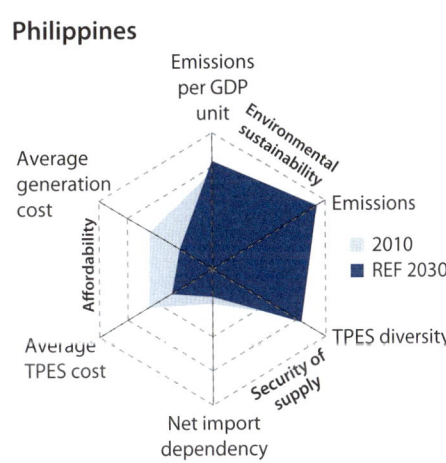

Thailand

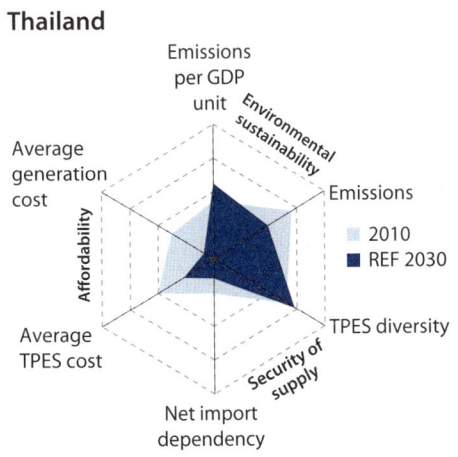

Vietnam

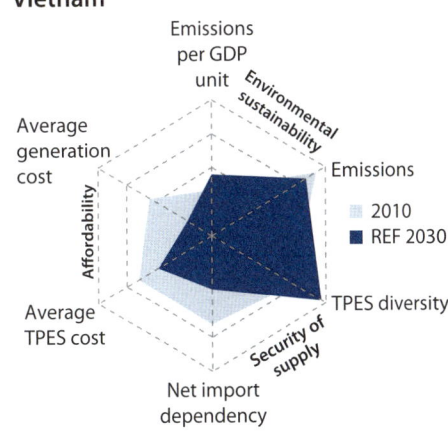

Source: Authors' calculations.

4

Sustainable Future: Improved Environment and Enhanced Security

Key messages: It is within the reach of the region's governments to simultaneously maintain economic growth, improve environmental sustainability, and enhance energy security. These goals require major domestic policy and institutional reforms, as well as transfers of substantial financial resources and low-carbon technologies from developed countries. This study found that it is technically and economically feasible to stabilize CO_2 emissions in East Asia by 2025. Increased energy efficiency is the backbone of this sustainable energy path. Major expansion of low-carbon technologies also is needed, and under the SED scenario, would meet half of the power demand in 2030. Carbon capture and storage (CCS) is expected to play an important role in future coal use in a carbon-constrained world, particularly beyond 2030. However, if CCS could become commercially available on a large scale by 2020, it could bring peaking time to 2021 and further reduce CO_2 emissions at a modest cost. To achieve the SED scenario requires a net additional investment of $80 billion per year from 2010 to 2030. After the initial three years, this investment would be offset by energy savings. However, financing the upfront investment costs is a major hurdle in developing countries.

4.1 Carbon emissions can peak in 2025

The SED scenario shows that it is technically and economically feasible for the region's carbon emissions to peak in 2025, and decline slightly decline thereafter, provided that there will be political will, institutional capacity, and transfer of financing and technologies from developed countries. CO_2 emissions of China and EAP5 countries would reach 9.2 Gt in 2030, 37 percent below the REF scenario (figure 4.1). China will be the main source of these emission reductions, reaching 7.7 Gt in 2030 (figure 4.2). These results are comparable to

Figure 4.1 Energy Efficiency and Low-Carbon Technologies Can Shift Region to a Sustainable Energy Path by 2030 *(Gt)*

Source: Authors' calculations.

Figure 4.2 CO$_2$ Emissions in China and EAP5 Countries under SED Scenario, 2010–30 *(MtCO$_2$)*

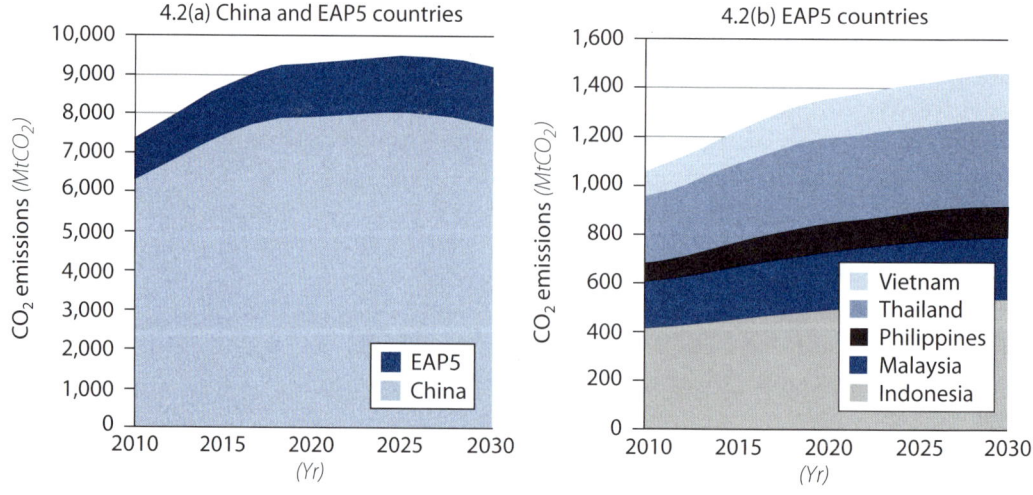

Source: Authors' calculations.

those of other leading international and Chinese studies.[28] Under the SED scenario, by 2030 in EAP5 countries, carbon emissions would reach 0.5 Gt in Indonesia, 0.4 Gt in Thailand, 0.3 Gt in Malaysia, 0.2 Gt in Vietnam, and 0.1 Gt in the Philippines (figure 4.2).

Energy efficiency contributes to more than half of the emission gap between the SED and REF scenarios by 2030 (figure 4.1).

---

28. McKinsey Global Institute 2008; China Energy Research Institute 2009; Clarke and others 2009; Riahi and others 2007; IIASA 2009.

Low-carbon power generation technologies (renewable energy and nuclear power) meet the remaining gap.

4.2 Local environment and energy security are improved

Under the SED scenario, energy efficiency and low-carbon technologies would lead to a 50 percent reduction in local environmental damage costs in 2030, compared to the REF scenario (figure 4.3). The benefits would be particularly significant for China in absolute terms, given its current reliance on coal, the most polluting fuel.

These measures also will improve security of supply by increasing fuel diversity and reducing imports. Under the SED scenario, all countries will improve their fuel diversity as they switch from fossil to low-carbon fuels and achieve a more balanced energy mix (figure 4.4). China shows the largest improvement with a nearly 45 percent increase in the IS Index, due mainly to the switch from coal to low-carbon fuels by 2030.

Under the SED scenario, compared to the REF scenario, the lower energy demand arising from energy efficiency measures would reduce the region's vulnerability to imports by 2030. The share of imported energy would fall significantly for all East Asia countries (figure 4.5). Under SED, the region would change from net imports of 22 percent under REF to net exports of 5 percent. This would be a very significant turn-around in the region's aggregate imports vulnerability. By 2030, under SED, Vietnam and Malaysia

Figure 4.3 Under SED Scenario, Local Environmental Damage Costs Would Be Halved, 2010–30 *(US$bil)*

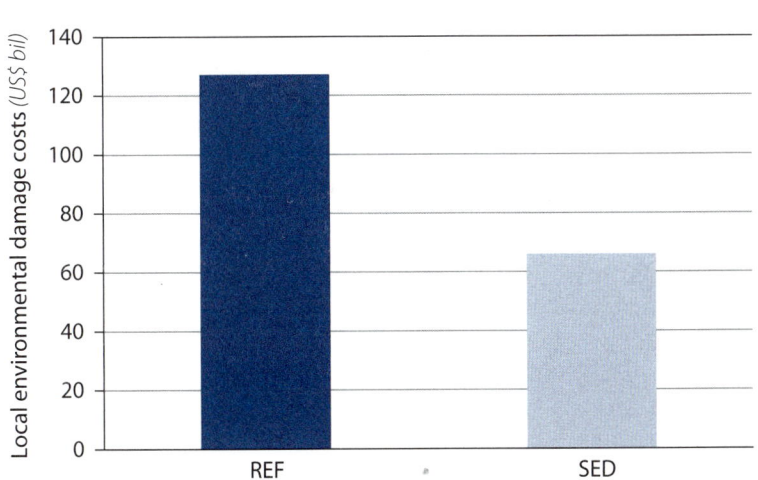

Source: Authors' calculations.

Figure 4.4 Fuel Diversity Improves across East Asian Countries, 2010–30 *(IS Index)*

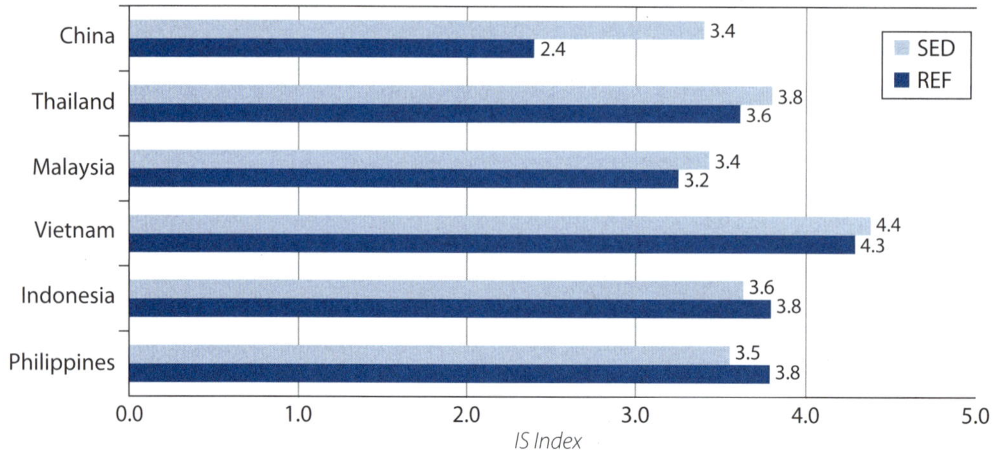

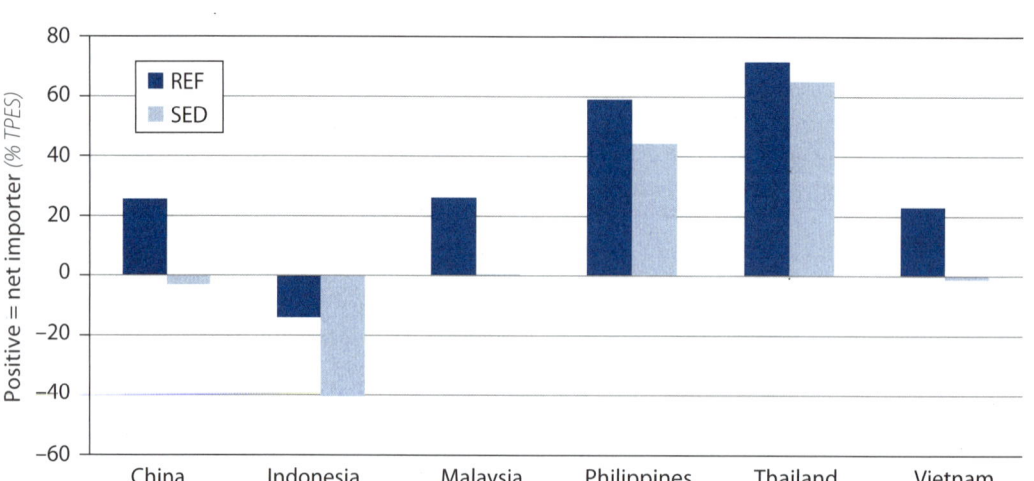

Source: Authors' calculations.

would switch from being net energy importers under REF to net exporters. Moving from the REF to the SED scenario would reduce Chinese oil imports by more than 240 million tons of oil equivalent (Mtoe), or 38 percent, by 2030.

In addition, adding renewable energy to the utility portfolio would hedge against fossil fuel price volatility risks. Increasing fuel prices by 20 percent would increase the costs of generation by 16 percent for gas and 6 percent for coal, while leaving renewable energy practically untouched (WEF 2009).

However, improving environmental sustainability and enhancing energy security also has trade-offs. Compared to the REF scenario,

Figure 4.5 East Asian Countries Reduce Their Reliance on Imports, 2010–30 *(% TPES)*

Source: Authors' calculations.

the SED scenario assumes more gas consumption in gas-producing countries including Indonesia, Malaysia, and Vietnam; and a higher share of nuclear power in the generation mix in China. These two increases could expose countries to risks of gas price volatility, nuclear reactor safety, and nuclear weapons proliferation.

4.3 Energy efficiency and low-carbon technologies make the difference

This study identified 2 main measures that would achieve the emission reductions in East Asia under the SED scenario by 2030:

- Energy efficiency on both supply and demand sides

- Low-carbon technologies: mostly renewable energy in China and EAP5, and nuclear power expansion only in China.

In the short term, natural gas could serve as an interim solution to shift from coal before major cost reductions and technology breakthroughs are achieved for renewable energy. Over the long term (beyond 2030), all leading global climate and energy models suggest that carbon capture and storage will be essential to achieve the deep emission cuts required to stabilize the climate, particularly in China, in which coal dominates.[29] Sensitivity analyses in this study show that if CCS could become commercially available on a large scale by 2020, it would play an important role in advancing the peaking time from 2025 to 2021 and further reducing CO_2 emissions by 5 percent in 2030, with a total additional financial cost of 1 percent,[30] compared to the SED scenario. Finally, the role of nuclear power is more flexible, but reducing its role would require an increase in fossil-based carbon capture and storage and renewables.[31]

Energy Efficiency

Energy efficiency presents the largest emission reduction source (figure 4.1). East Asian countries are much more energy intensive than industrialized countries. They, therefore, have a huge energy savings potential (figure 4.6).

Under the SED scenario, China would need to reduce energy intensity by 4.3 percent annually over the next 2 decades, compared to its record of a 3.4 percent reduction per year over the past decade

---

29. Knopf and others 2010; Rao and others 2008.
30. While CCS and solar PV technologies would push up the capital investments by 5% in the power sector, the total financial costs (including capital, operations, and fuel costs) in all sectors would increase by 1%.
31. Rao and others 2008; Calvin and others 2010; Knopf and others 2010.

Figure 4.6 East Asian Countries Are Energy Intensive *(toe/$MGDP 2008)*

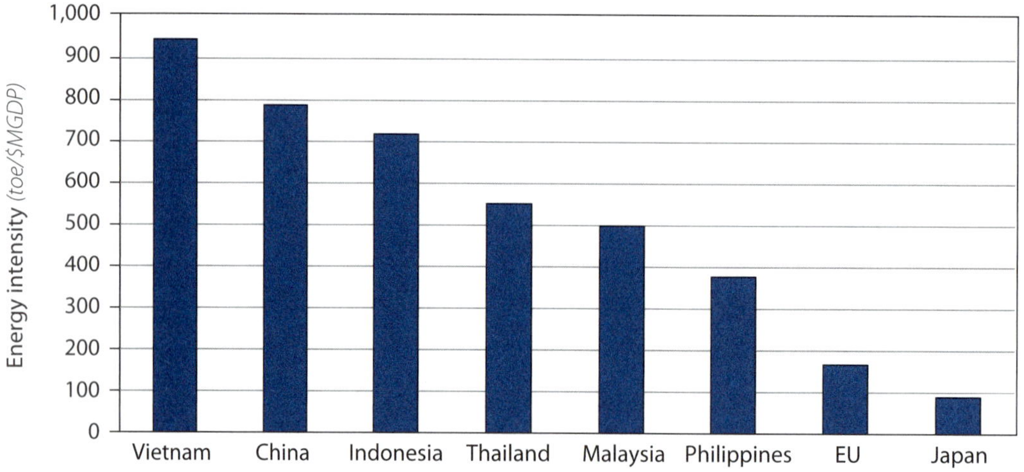

Source: Authors based on data from IEA 2008c.

and the current government's target of 4.2 percent per year. This energy-saving potential under the SED scenario is comparable to the assumptions made in other leading international and Chinese studies.[32] The energy intensity reduction under the SED is a daunting goal, given that China is at a developmental stage in which energy-intensive industries, driven by demand from domestic and export production, dominate the economy. For example, the sharp rise of the share of heavy industry in China since 2000, which is driven by strong demand from domestic and international markets, is the main reason for nearly doubled emissions from 2000 to 2005, despite the fall in energy intensity within industrial subsectors.

Hence, economic structural change is perhaps the single largest contributing factor to this reduction. In the past, economic structural change has contributed 45 percent–69 percent to energy intensity reduction in China. However, despite government's efforts during the 11th Five-Year Plan period, China has not made much progress in changing its economic structure (World Bank 2008b). Analysis shows that if the share of services in China (40 percent) were to reach that of the India (54 percent) or the US levels (76 percent), China's energy intensity would drop 22 percent and 31 percent, respectively (Lin and others 2007). In addition, the export-led growth model has resulted in exports responsible for approximately one-third of Chinese emissions of GHG in 2005 (Weber and others 2008). Therefore, rebalancing its economic structure is essential for China to achieve the full potential of energy efficiency improvements and ensure sustainable economic development.

---

32. ERI, APERC, IEA, McKinsey Global Institute, Stockholm Environment Institute, and Tyndall Centre.

Figure 4.7 SED Scenario Requires That Energy Intensity in East Asian Countries Decline Dramatically, 2007–30 *(toe/$M 2008 GDP)*

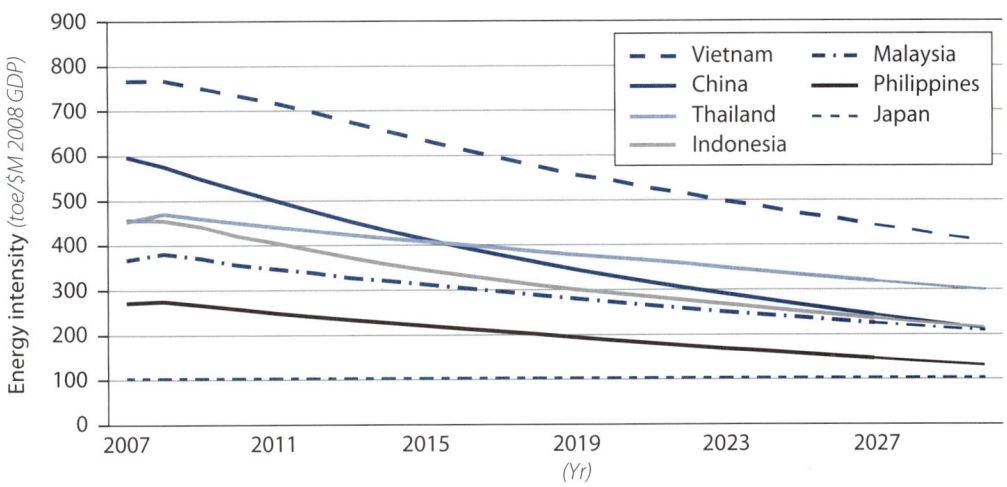

Source: Authors' calculations.

Under the SED scenario, EAP5 countries also would need to substantially reduce their energy intensities until 2030. The reduction would be by 3.3 percent per year for Indonesia, 3.1 percent for the Philippines, 2.8 for Vietnam, 2.5 percent for Malaysia, and 1.8 percent for Thailand (figure 4.7). For China and EAP5 countries to achieve such ambitious targets requires major policy and institutional reforms. If these reforms are in place, energy efficiency could reduce regional energy demand by more than 20 percent by 2030 by moving from the REF to the SED scenario (figure 4.8).

Figure 4.8 Primary Energy Supply for China and EAP5 Countries under SED Scenario, 2010–30 *(Mtoe)*

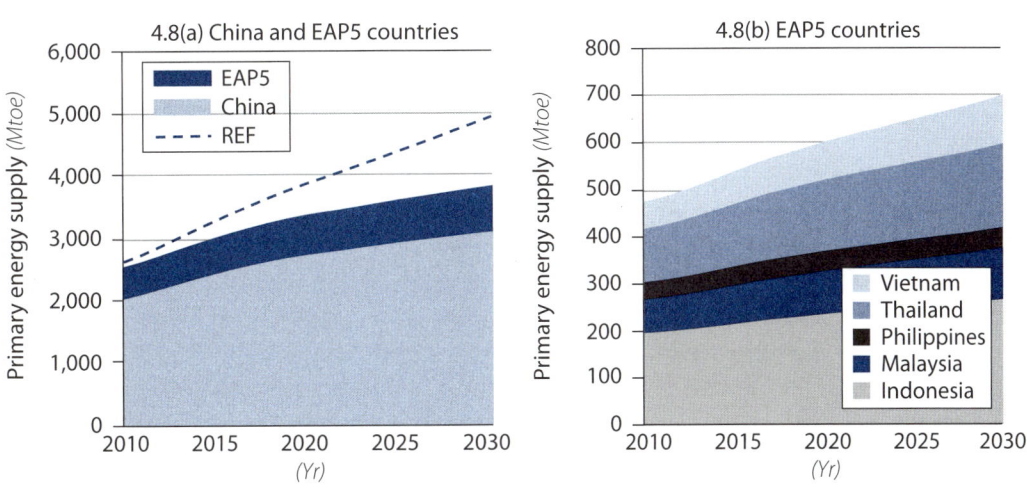

Source: Authors' calculations.

Box 4.1 Energy Efficiency Faces Many Market Barriers and Failures

- *Low or underpriced energy.* Low energy prices undermine incentives to save energy.
- *Regulatory failures.* Consumers who receive unmetered heat lack the incentive to adjust temperatures, and utility rate-setting can reward inefficiency.
- *No institutional champion and weak institutional capacity.* Energy-efficiency measures are fragmented. Without an institutional champion to coordinate and promote energy efficiency, it becomes nobody's priority. Moreover, energy-efficiency service providers are few, and their capacity cannot be established overnight.
- *Absent or misplaced incentives.* Utilities make more profit by generating and selling more electricity than by saving energy. For most consumers, the cost of energy is small relative to other expenditures. Because tenants typically pay energy bills, landlords have little or no incentive to spend on efficient appliances or insulation.
- *Consumer preferences.* Consumer decisions to purchase vehicles usually are based on size, speed, and appearance rather than on efficiency.
- *Higher upfront costs.* Many efficient products have higher upfront costs. Individual consumers usually demand very short payback times and are unwilling to pay higher upfront costs. Preferences aside, low-income customers may not be able to afford efficient products.
- *Financing barriers and high transaction costs.* Many energy-efficiency projects have difficulty obtaining financing. Financial institutions usually are not familiar with or interested in energy efficiency, because of the small size of the deal, high transaction costs, and high perceived risks. Many energy service companies lack collateral.
- *Limited awareness and information.* Consumers have limited information on energy-efficiency costs, benefits, and technologies. Firms are unwilling to pay for energy audits that inform them of savings.

Source: World Bank 2009a.

However, to realize this large potential of energy savings in East Asia is a major challenge and requires fundamental reforms. Energy efficiency improvements require well-functioning institutions. In many East Asian countries, governments find it politically difficult to raise energy prices, but low energy prices discourage energy conservation. Small-scale, fragmented energy-efficiency measures involving multiple stakeholders and tens of millions of individual decisionmakers clearly are more complex than large-scale, supply-side options. Energy-efficiency investments need cash up front, but future savings are less tangible, making such investment perceived to be risky compared with asset-based energy-supply deals.

Figure 4.9 Sensitivity Analyses Show Energy Efficiency Is Most Important Emission Reduction Option, While New Technologies Can Further Reduce Emissions, 2007–30 *(MtCO$_2$)*

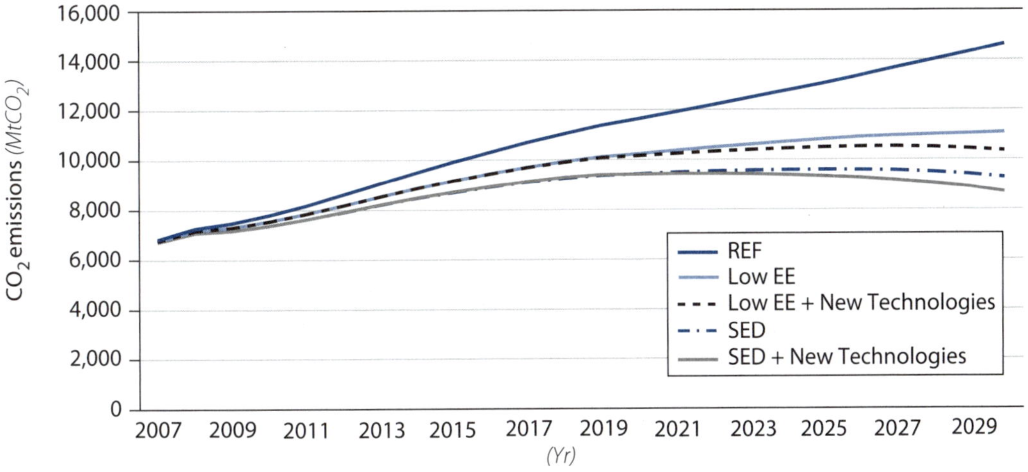

Source: Authors' calculations.

Many market failures and barriers to energy efficiency exist. Tackling them requires policies and interventions, but these entail additional costs (box 4.1).[33] Another concern is the rebound effect. Acquiring efficient equipment lowers energy bills, so consumers tend to increase energy consumption, eroding some of the energy reductions. However, the rebound is small to moderate, with long-run effects of 10 percent–30 percent for personal transport and space heating and cooling (Sorrell 2008), and price signals can mitigate it.

Given these challenges to achieve the ambitious energy efficiency assumptions in the SED scenario, this study conducted a sensitivity analysis to evaluate the impacts of an alternative scenario: what if the energy efficiency improvements were halfway between what is assumed under the REF and SED scenarios? The analysis shows that, compared to the SED scenario, in the low energy efficiency scenario, CO$_2$ emissions would peak by 2030 (a 5-year delay compared to SED) and increase by 20 percent in 2030, with a total additional financial cost of 4 percent (figure 4.9). This study also conducted a sensitivity analysis to compare the trade-offs between energy efficiency and new technologies. In this low energy efficiency and new technology scenario, energy efficiency improvements

---

33. For example, the transaction costs of mass distribution of compact fluorescent lamps (CFLs) are estimated at approximately $1 for distribution, awareness and promotion, monitoring and verification, and testing, which adds to the $1 hardware cost per CFL through bulk procurement.

would fall halfway between the assumptions of REF and SED scenarios and would include aggressive deployment of CCS and solar technologies. Compared to the SED scenario, in the low EE and new technology scenario, CO_2 emissions would peak at 2027 and increase by 12 percent, with a total additional financial cost of 6 percent. These results (table 4.1) demonstrate that energy efficiency is the most important and cost-effective emission reduction option and cannot be replaced by new technologies.

Industrial energy efficiency. In China, Indonesia, Thailand, and Vietnam, the industrial sector offers the single largest energy savings (figure 4.10). Key opportunities include improving the efficiency of energy-intensive equipment, such as motors and boilers, and of energy-intensive industries, such as iron and steel, cement, chemicals, and petrochemicals. Waste heat recovery and combined heat and power are among the most cost-effective measures. For example, China produces one-half of the world's cement and one-third of global steel. The application of the best available technologies in the cement and iron and steel industries could reduce CO_2 emissions by 250 Mton and 150 Mton, respectively. These amounts together account for 7 percent of the country's emissions (IEA 2008c). Indonesia has a significant potential to make cost-effective energy efficiency improvements in four manufacturing sectors: cement, textiles, basic metals, and food. These four account for half of the country's industrial emissions (World Bank 2009e)

Strategies to improve industrial energy efficiency differ by country. In China, improving energy efficiency requires not only embodying the most efficient technologies in new capital stocks, but also retrofitting and retiring existing inefficient technologies. Most of the new energy-intensive plants already have adopted state-of-art technologies. Nevertheless, the average industrial energy intensities are still far below the international benchmarks, due largely to the

Table 4.1 Sensitivity Analyses Results

	Peaking year	Emissions in 2030 (Gt)	Financial costs ($USbil/yr)	Investment cost ($USbil/yr)
REF scenario	—	14.6	1,167	100
SED scenario	2025	9.2	1,081	180
SED + new technologies scenario	2021	8.7	1,096	195
Low energy efficiency scenario	—	11.1	1,129	160
Low energy efficiency + new technologies scenario	2027	10.3	1,141	170

Source: Authors' calculations.
Note: Financial costs include capital investment, O&M, and fuel costs.

Figure 4.10 Industry Dominates Energy Demand in East Asian Countries *(%)*

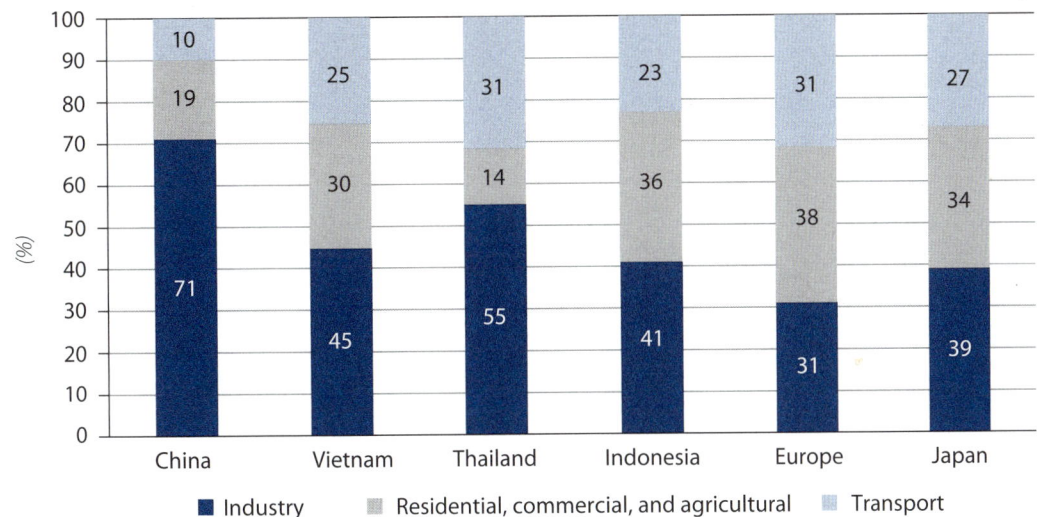

Source: Authors based on data from IEA 2008d.

existing inefficient energy stocks. Therefore, retrofitting and retiring existing inefficient plants are critical. In contrast, Vietnam's priorities should focus on incorporating efficient technologies in the new industrial capacity built over the next decade. This new capacity alone will produce more than all of Vietnam's industry today. At the same time, cutting energy waste in existing industry could save an additional 25 percent–30 percent of energy (World Bank 2009f).

Energy efficiency in buildings and appliances. The second largest energy saving potentials in most East Asian countries come from buildings in the residential, commercial, and public service sectors. Buildings consume nearly 40 percent of the world's final energy.[34] Although East Asia's emissions currently are dominated by the power and industrial sectors, over the next 20 years, the transport and building sectors are expected to grow more rapidly due to unprecedented urbanization. It is projected that 300 million people will migrate to China's urban areas over the next 2 decades, and that two-thirds of China's 1.5 billion people will live in urban areas by 2030. In the past decade, China already underwent one of the biggest building booms in history, and the total floor space is expected to more than double from 2005 to 2030 (figure 4.11), stoking consumer demand for energy (McKinsey Global Institute 2008).

34. IEA 2008b; Worldwatch Institute 2009.

Figure 4.11 China Will Double Building Floor Areas and Increase Its Vehicle Fleet 10-Fold, 2005–30

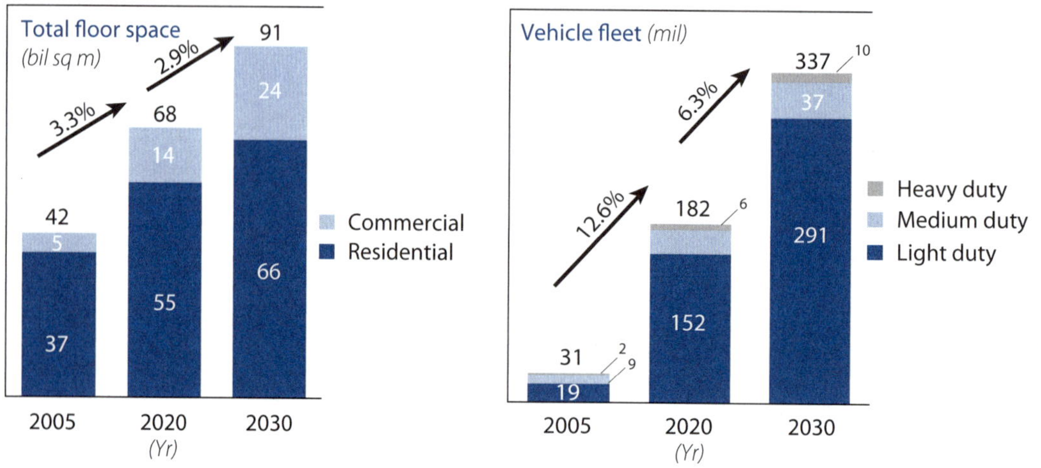

Total floor space *(bil sq m)*

2.9%

3.3%

	2005	2020	2030
Commercial	5	14	24
Residential	37	55	66
Total	42	68	91

Vehicle fleet *(mil)*

6.3%

12.6%

	2005	2020	2030
Heavy duty	2	6	10
Medium duty	9		37
Light duty	19	152	291
Total	31	182	337

Source: McKinsey Global Institute 2008.

Buildings present one of the most cost-effective mitigation options. Opportunities to improve energy efficiency lie in the building envelope (roof, walls, windows, doors, and insulation), in space and water heating, and in appliances (lighting, air conditioning, and refrigeration). Studies find that, when evaluated on a life-cycle basis, existing energy-efficiency technologies can cost-effectively save 30 percent–40 percent of energy use in new buildings.[35]

Introducing energy efficiency in new buildings is more cost effective than retrofitting existing buildings. Nearly half of the urban building stocks projected to exist in China in 2030 have yet to be built. The current space-heating technology used in Chinese buildings consumes 50 percent–100 percent more energy than that used in Western Europe. Making buildings in China more energy efficient would add 10 percent to construction costs but would save more than 50 percent on energy costs over the life of the building (Shalizi and Lecocq 2008). Integrated zero-emission building designs, which combine energy efficiency measures with on-site power and heat from solar and biomass, are technically and economically feasible—and their costs are falling (Brown and others 2005).

In EAP5 countries, the parallel household study found that the total electricity consumed by the residential sector is expected to more than double from 2006 to 2030 (appendix 2). Most of this growth will be driven by a sharp increase in ownership of

---

35. Brown and others 2005; Burton and others 2008. A comprehensive review of empirical experience based on 146 green buildings in 10 countries concluded that green buildings cost on average approximately 2% more to build than conventional buildings but could reduce energy use by a median of 33% (Kats 2008).

Figure 4.12 Growth in Residential Electricity Consumption Is Driven by Air Conditioners, TVs, and Refrigerators, 2006 *(GWh/yr)*

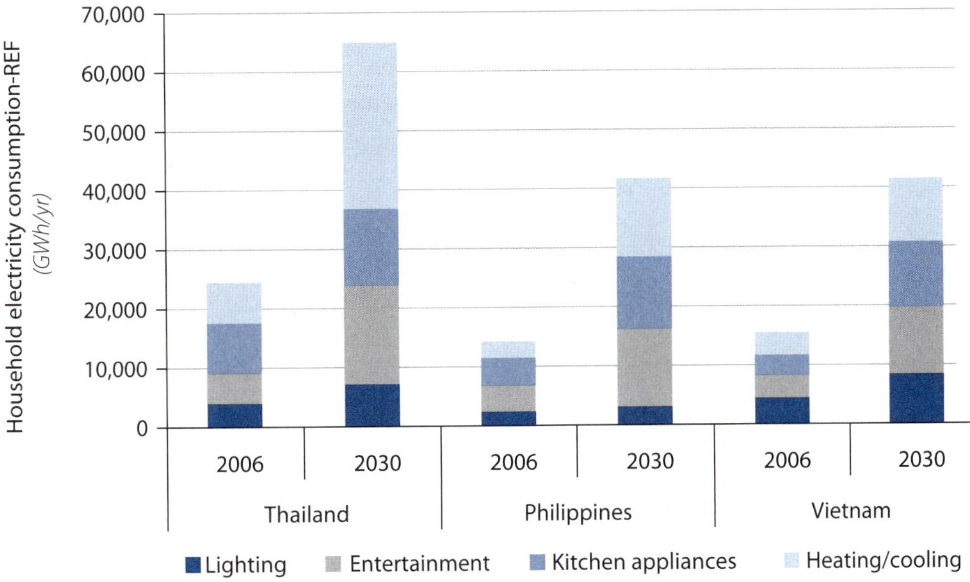

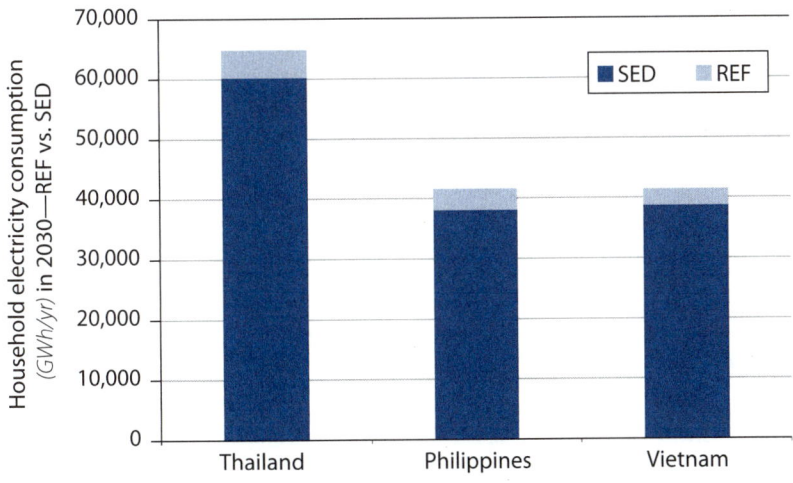

Source: Authors' calculations.

high-electricity-consuming appliances: air conditioners, televisions, and refrigerators (figure 4.12). Introducing efficiency standards to new appliances based on the most efficient models in the market today would result in an average annual savings of 8 percent in energy consumption compared to the baseline in 2030 (figure 4.13).

Figure 4.13 Improvements in Appliance Efficiency Can Save 8 Percent of Annual Energy Consumption Compared to the REF Scenario in 2030 *(GWh/yr)*

Source: Authors' calculations.

Figure 4.14 Fuel Economy Standards and Public Transport Infrastructure Make Biggest Difference in Reducing Transport Fuel Consumption, 2007–20 *(%)*

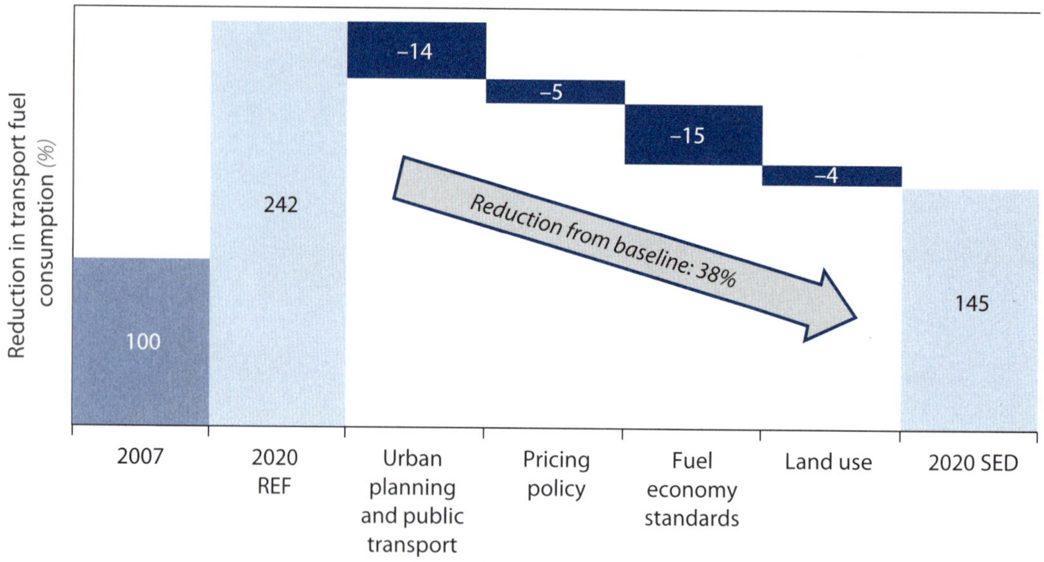

Source: Authors' calculations.

Transport energy efficiency. In 2009, for the first time, China had the world's largest automobile sales, surpassing the United States. By 2030, over 330 million vehicles will be on China's roads (figure 4.11). China therefore will need a strategy to avoid the road to oil dependence that developed markets have followed (McKinsey Global Institute 2008).

This study found that clean vehicles (hybrid and electric vehicles) that meet higher fuel economy standards, coupled with public transport and urban planning, make the biggest difference in containing energy consumption in urban transport (appendix 1). The parallel transport study shows that, given three conditions, energy use and emissions in the transport sector could be reduced by 38 percent from the baseline by 2020. The three conditions are that the transport master plans (urban planning and public transport) are fully met; future EU fuel efficiency standards are applied to new vehicles (requires highly efficient and electric vehicles); and road pricing (congestion charges, parking fees, and fuel taxes) measures are taken (figure 4.14). Achieving and adhering to these three conditions would reduce oil imports by 30 percent in most East Asian countries. Cities including Manila, Ho Chi Minh City, Chengdu, and Beijing stand to benefit the most if all these actions are taken, because their energy consumption due to transport could go down by 50 percent.

In the near to medium term, improving vehicle fuel efficiency is the most cost-effective means of cutting emissions in the transport sector. Plug-in hybrids offer a potential near-term option as a means

Figure 4.15 Potential Savings Are Significant but Vary by City and Strategy *(%)*

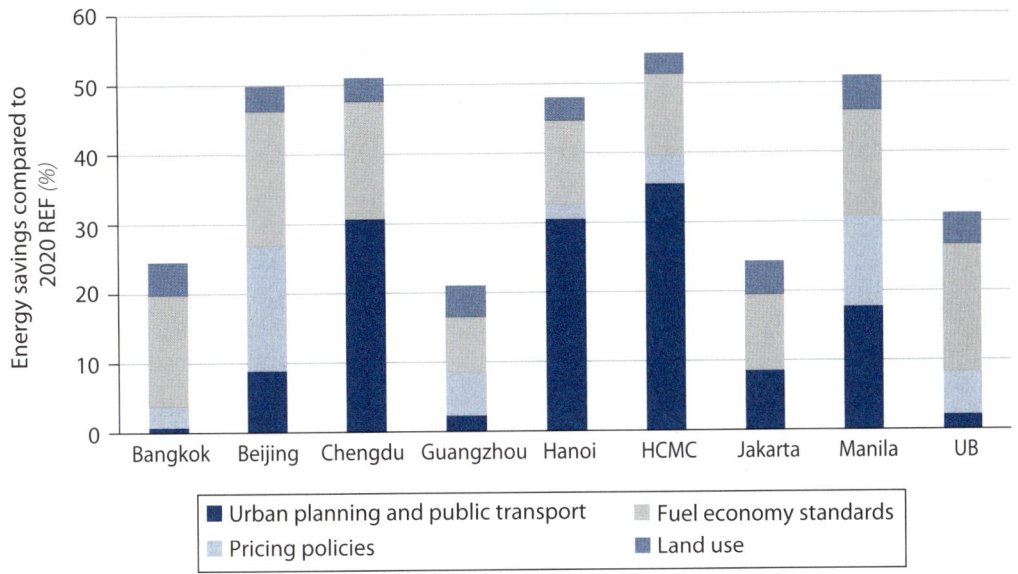

Source: Authors' calculations.

to transition to full electric vehicles.[36] When running on electricity generated from renewable energy, these hybrids emit 65 percent less CO_2 than a gasoline-powered car (NRDC 2007). However, both plug-in hybrids and electric vehicles increase electricity consumption, so the net emission reductions depend on the electricity source. In addition, significant improvements and cost reductions in energy storage technology are required.

In addition, smart urban planning can substantially reduce energy demand and CO_2 emissions. Smart urban planning means higher density and more spatial compactness; and more mixed-use urban design, which enables growth near city centers and transit corridors to prevent urban sprawl. Compact urban design reduces the vehicle kilometers traveled and enables relying on district and integrated energy systems for heating.[37] Furthermore, modal shifts to mass transit can be costly. However, these costs are justified by mass transit's large development co-benefits of time savings in traffic, less congestion, and better public health from reduced local air pollution.

Strategies to contain transport fuel consumption vary by city (figure 4.15). For established megacities such as Beijing, Bangkok, and Manila, higher fuel economy standards and clean vehicles, coupled with

36. IEA 2008b. Plug-in hybrid vehicles combine batteries with smaller internal combustion engines, which enable them to travel part-time on electricity provided by the grid through recharging at night.

37. Brown and others 2005; ETAAC 2008.

appropriate road pricing and fuel taxes, would make the biggest difference. For medium-sizes cities such as Chengdu, Hanoi, and Ulaanbaatar, urban planning and public transport would be most effective.

Low-Carbon Technologies

Energy efficiency measures alone will not be sufficient for CO_2 emissions to peak in 2025. To achieve the emission reduction under the SED scenario, also requires a three-fold increase in low-carbon technologies from the current levels by 2030. Under the SED scenario, low-carbon technologies would meet half of the power demand by 2030, a big step up from the current 17 percent, (figure 4.16). The share of coal in the power mix would decline from 70 percent under the REF scenario to 37 percent under the SED scenario by 2030.

Renewable energy. Under the SED scenario, renewable energy would meet 40 percent of the power demand by 2030. The renewables would come largely from hydro, wind, and biomass in China; hydro, biomass, and geothermal in Indonesia; geothermal and hydro in the Philippines; imported hydro and biomass in Thailand; and hydro in Vietnam.

Figure 4.16 Under the SED Scenario, Power Generation Shifts Dramatically from Coal to Renewable Energy and Nuclear Power *(%)*

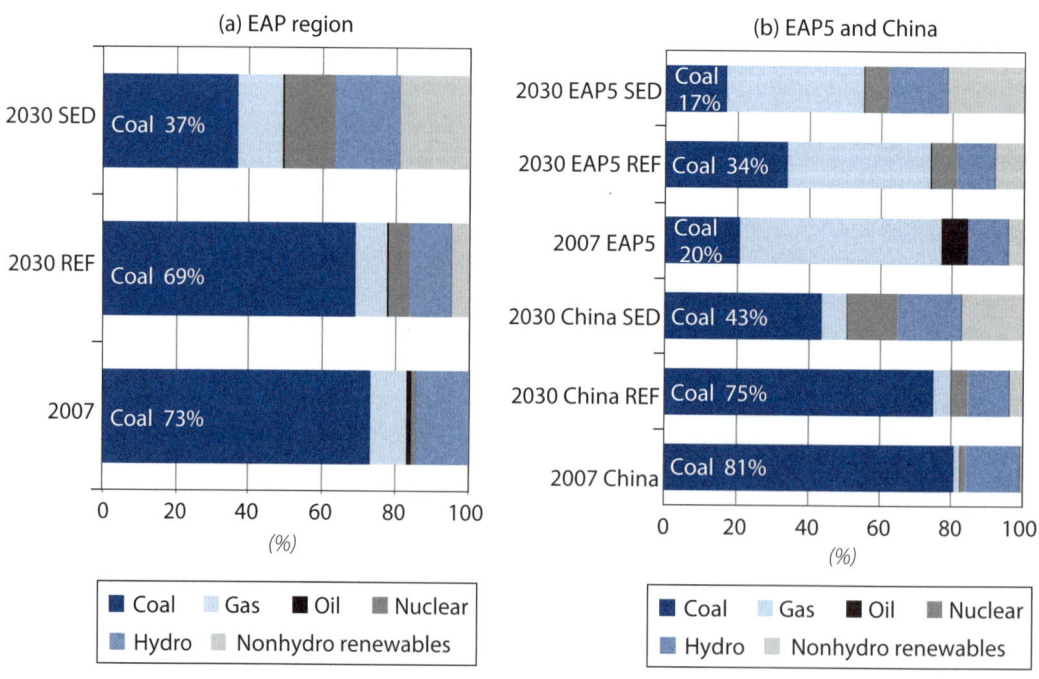

Source: Authors' calculations.

Table 4.2 East Asian Countries Have Rich Renewable Energy Resources *(GW power capacity)*

	China	Indonesia	Thailand	Philippines	Vietnam
Hydro	400	75	0.7	3.0	22
Wind	380	9	1.6	5.5	2
Biomass	60	50	4.4	0.5	1
Geothermal		27		3.0	1.4
Total	840	160	6.7	12.0	26

Source: China data from National Development and Reform Commission, Renewable Energy in China 2008; Indonesia data from Ministry of Mines and Energy, cited in IEA *Energy Policy Review of Indonesia 2008*; Thailand data from Department of Alternative Energy Development and Efficiency, Ministry of Energy, July 2009; the Philippines data from Department of Energy; and Vietnam data from Nguyen and others 2009.
Note: Hydro includes both large and small hydro; wind includes only on-shore wind.

The potential for expanding renewable energy depends largely on resource availability. Wind, hydropower, and geothermal power are limited by availability of suitable sites. Biomass is constrained by competition from land and water for food and forests. Solar power is the most abundant energy source on earth but is still costly. Renewable energy resource potential is large in the region, particularly in China, Indonesia, the Philippines, and Vietnam (table 4.2). Specifically, China, Indonesia, and Vietnam have rich hydro resources; and Indonesia has the world's largest geothermal resources.

Regional hydropower trade could provide the least-cost energy supply with zero carbon emissions. Lao DPR, for example, is rich in hydro power resources, but the size of its power market and economic fundamentals are not sufficient to justify and enable development of those resources on their own. On other hand, Thailand has limited renewable energy resources, and Cambodia relies heavily on diesel for power generation with a high cost of 25–30 cents/kWh. The evident solution is regional power trade among neighboring countries. The main effect of power trade is to support the development of a higher number of large-scale hydro schemes that would not otherwise be viable at the national level.

Costs of renewable energy have declined dramatically over the past two decades. Small hydro is now cost-competitive with coal. Wind, geothermal, and biomass co-generation from wastes can be economically viable compared to the costs of coal-fired power plants plus local and global environmental external costs (figure 4.17).[38] With rising fossil-fuel prices, the cost gap is closing. Biomass,

38. The costs of wind, geothermal, and hydro vary greatly depending on resources and sites.

Figure 4.17 Geothermal Power in Indonesia Is Economically Competitive with Coal-Based Power *(cents/kWh)*

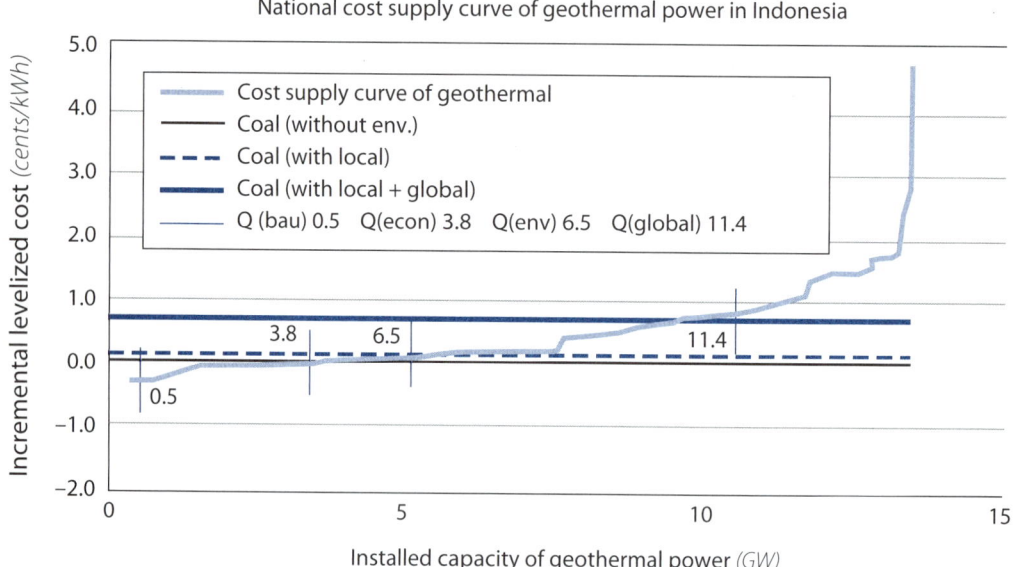

Source: World Bank staff calculations.

geothermal power, and hydropower can provide base load power; and solar and wind power are intermittent.

Solar energy is still costly. However, over the next few years, costs are expected to decline rapidly along the learning curve due to technology breakthroughs and economy of scale.[39] To test the robustness of cost reduction, this study conducted a sensitivity analysis to examine the impacts of large-scale penetration of solar technology compared to the SED scenario (figure 4.9).

A large share of nondispatchable (intermittent) resources in the power grid may affect the reliability of electricity supply. However, this unreliability can be addressed in a variety of ways: through hydropower (including pumped storage), load management, energy storage facilities, interconnections with other countries, and smart grids (IEA 2008b). Smart grids can enhance the reliability of electricity networks when incorporating intermittent renewable energy and distributed generation. High-voltage direct-current lines can make possible long-range transmission with low transmission losses. This technology reduces the problem of evacuating large amounts of renewable energy from locations rich in renewable resources, which often are far from consumption centers. Furthermore, additional cost reduction and performance improvement of energy storage will be needed for large-scale deployment of solar

39. Deutsche Bank Advisors 2008 (projected PV cost reductions).

and wind power and electric vehicles. Therefore, while the required magnitude of renewable energy is vast and technical challenges still significant, cost-effective energy conversion and delivery to consumers are increasingly feasible. For example, wind already accounts for 20 percent of Denmark's power production.

Nuclear power. The expansion of nuclear power from REF to SED scenarios in this study comes solely from China, thanks to the government's aggressive policies in this regard. Currently, among the East Asian countries, only China has commercial nuclear power generation. China is expected to rapidly expand its nuclear power capacity. As a matter of fact, it recently changed its policy target for nuclear power from 40 GW to 70 GW by 2020. The REF scenario is based on the existing government plan announced a few years ago, which assumed that the installed capacity of nuclear power in China would increase from today's 7 GW to 65 GW by 2030. The SED scenario assumes a more than double—140 GW—installed capacity by 2030 extrapolated from the latest government's target.

Malaysia, Thailand, and Vietnam also have firm plans to introduce nuclear power. However, building the first new nuclear power plant will be a long process, and none is expected to be in operation before 2020. In these countries, the SED scenario keeps the nuclear power capacity at the same level, based on government's power development plans, as in the REF scenario.

Nuclear power is a significant option for mitigating climate change, but it is limited by four problems. They are: higher costs than coal-fired plants,[40] risks of nuclear weapons proliferation, uncertainties about waste management, and public concerns about reactor safety. Current international safeguards are inadequate to meet the security challenges of expanded nuclear deployment (MIT 2003). However, the next generation of nuclear reactor designs offer improved safety characteristics and better economics than the reactors currently in operation.

Nuclear power has large requirements for capital and highly trained personnel, with long lead times before it comes on line. These factors lower its potential to reduce carbon emissions in the short term. Planning, licensing, and constructing a single nuclear power plant typically take a decade or more. Due to the dearth of orders in recent decades, the world has limited capacity to manufacture many of the critical components of nuclear power plants. Rebuilding this capacity alone will take at least a decade (Worldwatch Institute 2008). For example, the world's first third-generation nuclear power plant in Finland is three years later than scheduled.

---

40. MIT 2003; Keystone Center 2007.

Similarly, the Sizewell B nuclear power plant in the UK, intended to be the first of a series of 6 plants, took 15 years until commissioning, of which 7 years were spent obtaining the necessary approvals.

Natural Gas

Under the SED scenario, natural gas is expected to play an important role in East Asia. Opportunities are greatest in the electricity generation sector. Southeast Asian countries, particularly Indonesia and Malaysia, have large gas reserves and resources. With favorable policies and pricing, over the next two decades, gas production and consumption could increase substantially and play an important role in meeting the region's growth in demand.

Increased natural gas production will require a mix of incentives to attract more investment in upstream exploration and production. International investment in East Asia requires countries to offer attractive and competitive fiscal terms as well as other supporting policies. Uncertainty over the amount of gas that will be allowed to be exported (to earn foreign currency at international market prices) could have a negative impact on upstream investors. East Asian governments may be able to attract more investment to their upstream sectors by reviewing their fiscal terms in light of international comparators and removing uncertainty on exportable volumes. With these measures, additional production could come from Indonesia and Vietnam.

Domestic gas price subsidies, if reflected in prices paid to producers, also will have a negative impact on long-term production. Gas price subsidies were high in China, Indonesia, and Malaysia in 2008 when international oil and gas prices were peaking. Since international oil prices have fallen (and gas prices even further), these subsidies have substantially disappeared. If and when oil and gas prices rise in the future, it is important that large subsidies are not reimposed.

In view of the remote locations of gas reserves in EAP5 countries and the limited gas interconnections in the region, liquefied natural gas (LNG) trade is becoming more important. Most EAP5 countries are building domestic LNG receiving/regasification terminals. Malaysia and Indonesia are the world's leading LNG exporters. Japan and Korea are the world's biggest LNG buyers, and rely exclusively on LNG imports for their gas supply (IEA 2009).

However, in China, fuel switching from coal to gas is more limited. Even with by far the fastest rate of growth of gas use compared to the EAP5 countries, China may do little better than double its share of gas in total energy from the current 3 percent by 2030.

Large gas reserves exist in Russia and the Central Asian republics. Pipelines would be needed to bring some of these supplies to East Asia. A first pipeline is under construction to deliver 40 bcm/yr of gas from Turkmenistan to China. Other pipelines from Russia are under consideration, such as from Sakhalin to Korea. If more pipeline supplies into East Asia are able to be developed, the construction of more links of the Trans-Asian Gas Pipeline project would bring benefits to other countries in the region.

Rising dependency on gas imports in East Asia will increase the risks to the security of supply, via both possible supply interruption and exposure to price volatility. The costs of natural-gas-fired power depends on gas prices, which in recent years have been highly volatile. Long-term contracting of gas imports has been the traditional route taken to manage these risks. Despite the recent interest and growth in short-term gas trades, the long-term contracting approach should continue. Another growing trend is for the larger national oil companies (NOCs) to buy into overseas oil and gas-producing assets. The large Chinese NOCs are by far the most active in this area and are likely to continue to acquire ownership in foreign reserves, with some of other East Asian NOCs aiming to follow suit. Although buying into overseas assets is an understandable response to growing imports requirements and recent high prices, it remains to be seen whether this strategy will actually mitigate the risks of physical shortages, or provide only a limited financial hedge against high future prices.

Efficient Coal-Fired Power Plants and Carbon Capture and Storage

East Asian countries have large potential to improve the efficiency of their coal-fired power plants, as most of these power plants remain below the international benchmarks (figure 4.18). A majority of the new coal-fired plants in China are equipped with the most efficient supercritical and ultrasupercritical technologies in the world, at no additional costs compared to the subcritical plants.[41] Over the last decade, China also increased the average efficiency of coal-fired power plants by 15 percent to an average of 34 percent (Zhang 2008). Over the last two years, a policy that requires closing small-scale coal-fired power plants and replacing them with large-scale efficient plants reduced annual CO_2 emissions by 60 million tons.

Most EAP5 countries still plan to adopt subcritical technologies for their coal expansion under the REF scenario, and should use the most efficient technologies in their new plants, whenever possible. However, the likely continued use of small coal plants in some East Asian countries—for island systems in Indonesia, Malaysia and the Philippines—makes using the most efficient technologies difficult. Supercritical and ultrasupercritical technologies currently are

---

41. Supercritical and ultrasupercritical plants use higher steam temperatures and pressures to achieve higher efficiency of 38%–40% and 40%–42% respectively, compared to large-scale subcritical power plants with an average efficiency of 35%–38%.

Figure 4.18 Coal Plant Efficiency in East Asia Is Improving but Remains below International Benchmarks, 1995–2005 *(%)*

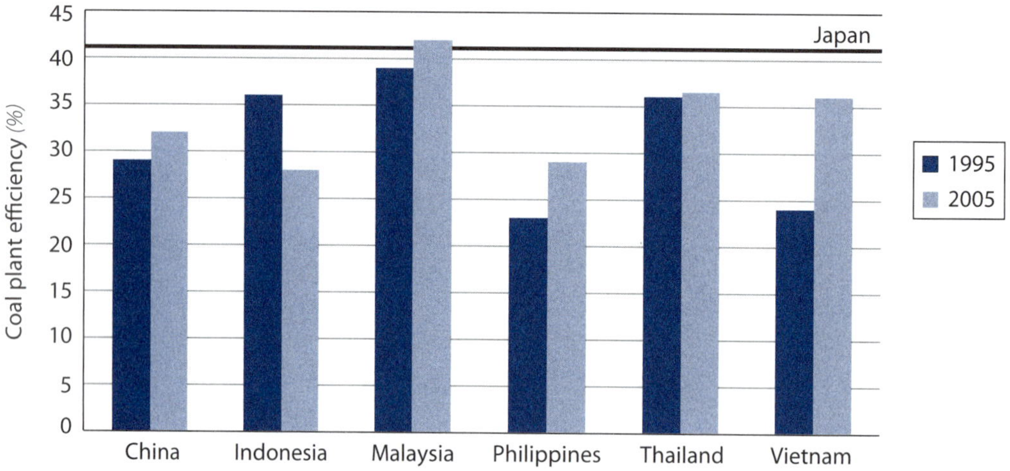

Source: Authors based on data from IEA 2008d.

available only in larger unit sizes due to scale economies and manufacturer choices.

The future use of coal in a carbon-constrained world will depend on widespread use of carbon capture and storage (CCS)—a promising technology but yet to be proven on a large scale (box 4.2). CCS technologies are costly, competitive with conventional coal only at a price of $50–$90/ton of CO_2 (IEA 2008b). CCS can significantly reduce the efficiency of power plants. Its potential in some countries may be limited by proximity to storage sites and possible leakage.

The near-term priority should be spurring large-scale demonstration projects to reduce costs and improve reliability. Four large-scale commercial CCS demonstration projects are in operation—in Sleipner (Norway), Weyburn (Canada-United States), Salah (Algeria), and Snohvit (Norway)—mostly from gas or coal gasification. Combined, these projects capture 4 million tons of CO_2 per year. In addition, capturing CO_2 from low-efficiency power plants is not economically viable, so new power plants should be built with highly efficient technologies for retrofitting with CCS later. Furthermore, legal and regulatory frameworks must be established to address CO_2 injection and long-term liabilities. Detailed assessments of potential carbon storage sites also are needed, particularly in developing countries. Without a massive international effort, resolving the entire chain of technical, legal, institutional, financial, and environmental issues around CCS could require a decade or more before applications go to scale (see chapter 5 for more details).

Over the medium term (beyond 2030), CCS is essential to "bend" the emission curve. Given the importance and uncertainty of CCS

Box 4.2 Carbon Capture and Storage

Carbon capture and storage (CCS) could reduce emissions from fossil fuels by 85 percent–95 percent and is critical in sustaining an important role for fossil fuels in a carbon-constrained world. The process involves three main steps:

1. *CO_2 capture from large stationary sources*, such as power plants or other industrial processes. Three main technology options exist for CO_2 capture: (a) pre-combustion (using integrated gasification combined cycle, or IGCC, technology); (b) post-combustion (using chemical absorption to separate CO_2 from the flue gas stream for existing coal- and gas-fired power plants); and (c) oxyfueling (using oxygen rather than air for combustion). CO_2 capture and pressurization require energy, reduce overall energy efficiency, and add cost. Typical efficiency losses today are 10–12 percentage points. Near-term opportunities with lower CO_2 capture costs are through industrial processes such as coal gasification, liquefaction, and fertilizer production.

2. *Transport to storage sites by pipelines*.

3. *Storage through injection of CO_2 into geological sites*. Early opportunities to lower costs exist at depleted oil and gas fields by enhancing oil and gas recovery, or at coal beds by enhancing coal bed methane recovery. However, for deep emission cuts, the most costly option of storage of CO_2 in deep saline formations and ocean also would be required.

technology, this study conducted a sensitivity analysis. The analysis shows that if CCS could become commercially available on a large scale by 2020,[42] it would advance the peaking time to 2021 (4 years earlier) and further reduce CO_2 emissions by 5 percent in 2030, with a total additional financial cost of 1 percent, compared to the SED scenario (see figure 4.9). This result demonstrates that, with a modest additional cost, the SED + CCS scenario could achieve an earlier CO_2 emissions peaking time at a lower emission peaking level.

Given the large domestic reserves, coal likely will remain an important energy source in China for decades. Carbon capture and

---

42. This sensitivity analysis represents the upper boundary of deployment of CCS and solar energy. It assumes that CCS technology will become commercially available on a large scale by 2020 and that the costs of solar energy will be significantly reduced. For example, it assumes that 25% of all coal-fired power capacity in China will be equipped with CCS by 2030.

storage are essential for China's economic growth within a carbon budget. The good news is that 90 percent of existing large stationary CO_2 point sources are within 100 miles (161 km) of a candidate deep geologic storage formation in China. The East and North regions have the most potential CCS sites, whereas the Northwest and Southwest have the fewest. The preliminary cost curve analysis suggests that the majority of emissions from China's large CO_2 point sources can be stored in large deep saline formations at estimated transport and storage costs of less than $10/t$CO_2$ (Dahowski and others 2008). Capture costs comprise 90 percent of total CCS costs.

4.4 Sustainable energy path is affordable, but financing is a major hurdle

Financing Is a Major Challenge

The SED scenario would need a total investment cost of $3.6 trillion in EAP5 countries plus China, or an average of $180 billion per year from 2010 to 2030. Of the annual average, $20 billion would go to thermal power, $55 billion to renewable energy, $20 billion to nuclear power, and $85 billion to energy efficiency. China would account for 85 percent of this investment.

Compared to the REF scenario, the SED scenario requires an additional average investment of $85 billion per year in energy efficiency in the power, industry, and transport sectors and $35 billion per year in low-carbon technologies ($25 billion in renewable energy and $10 billion in nuclear power) from 2010 to 2030 (figure 4.19). In the meantime, due to energy efficiency measures, the SED scenario avoided investments in thermal power plants of an average of $40 billion per year. The result is a net additional average capital investment of $80 billion per year from 2010 to 2030, or 0.8 percent of GDP in the region.

A formidable challenge is to mobilize sufficient financing to expand energy efficiency and capital-intensive low-carbon technologies. Financing has been a constraint in developing countries, resulting in under-investment in infrastructure as well as a bias toward energy choices with lower upfront capital costs, even when such choices eventually result in higher overall costs. Many clean energy investments have high upfront capital costs, followed later by savings in fuel costs. In fiscally constrained developing countries, these high upfront capital costs are a significant barrier to investment in low-carbon technologies. To overcome this barrier, concessional financing is required to cover the incremental costs and risks of clean energy solutions (chapter 5).

Figure 4.19 Financing the Sustainable Energy Path Is a Major Challenge *(US$bil)*

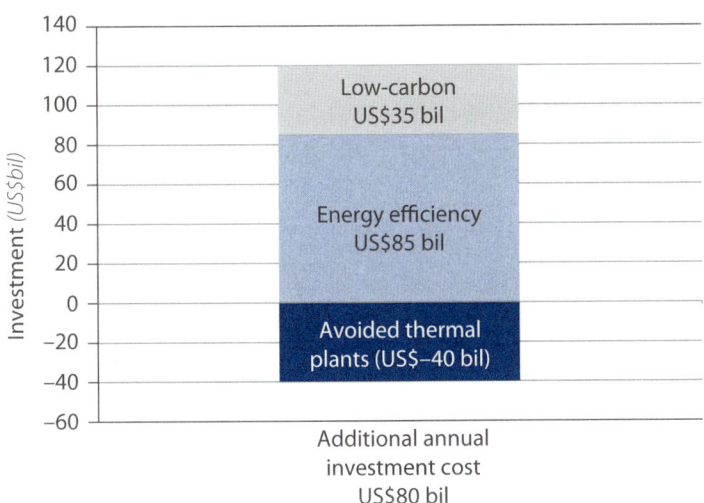

Source: Authors' calculations.

Additional Investments Can Be Offset by Energy Savings

1. Increasing energy efficiency will lead to net reductions in financial costs. On average, the investment on the demand side energy efficiency needed to save energy is considerably less than the investment on the supply side to increase the same amount of energy. This is most obvious in the electricity sector. Supply-side investment is typically 2–3 times as high as the cost of investment in energy efficiency to avoid electricity generation, transmission, and distribution expansion.

2. The reductions in fuel costs through energy savings can largely pay for the additional investment costs. At a discount rate of 10 percent, the annual net present value (NPV) of the fuel cost savings from 2010 to 2030 would amount to $145 billion, which is far greater than the annual NPV of the additional investment costs required, or $70 billion (figure 4.20). China would save $120 billion per year on fuel costs, or 1.5 percent of GDP. EAP5 countries would save a collective $25 billion per year on fuel costs, or 1.4 percent of GDP.

3. The cost savings are even larger when measured in economic terms. The annual NPV of economic cost savings from reduced environmental damage costs of local air pollution and lower CO_2 emission costs is estimated at $55 billion from 2010 to 2030 (figure 4.20).

Figure 4.20 Fuel Cost Savings and Lower Environmental Costs Are Much Larger Than Additional Investments
(US$bil)

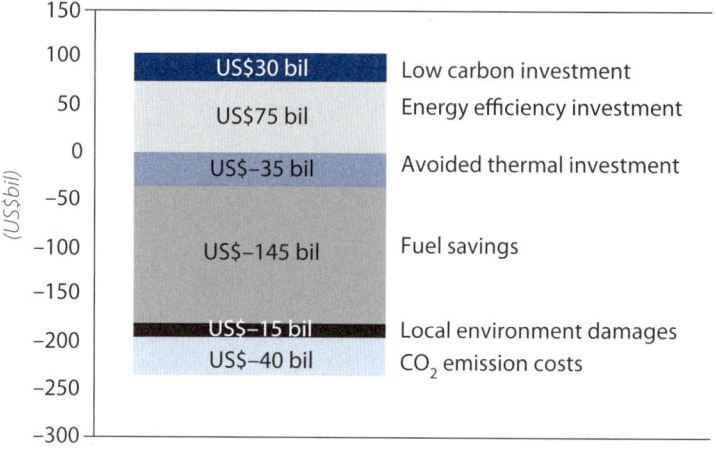

Source: Authors' calculations.

The energy savings can largely pay for this additional investment cost. While energy efficiency, many renewable energy technologies, and nuclear power require additional upfront investment, they can recoup part or all of that investment through lower energy spending from fuel savings in future years. Therefore, after the initial three years, the total financial costs of the SED scenario (including capital investment, O&M, and fuel costs) would not exceed those of the REF scenario.

In reality, achieving the energy efficiency potential depends on political will, policy environment, and institutional reforms. For example, several East Asian countries still have fossil fuel subsidies, and it will take strong political will to remove them. In addition, these financial costs in the energy sector do not capture the costs related to economic structural adjustments, which is perhaps the single largest contributing factor to energy efficiency improvement in China and possible subsidies to mitigate the impacts of higher energy prices on the poor.

Sustainable Energy Path Is Affordable

The impact on affordability depends on the change in energy costs and income levels. This study assesses the average level of income for each country, rather than a disaggregated analysis of household income and expenditure distributions due to limited data availability. Clear patterns emerge in the trends of energy costs compared to income levels.

Figure 4.21 Environmental Sustainability, Energy Security, and Costs of SED Scenario, 2030

China

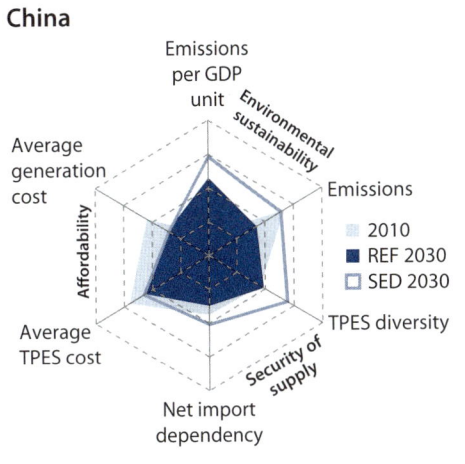

Indonesia

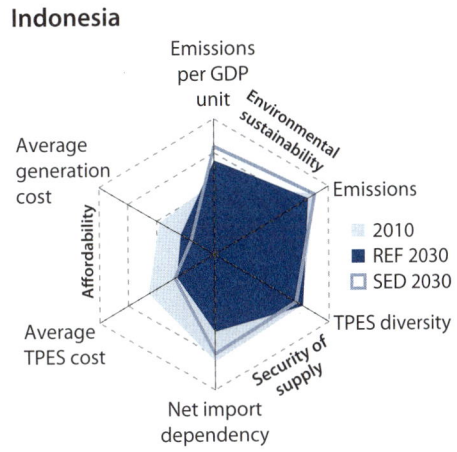

Malaysia

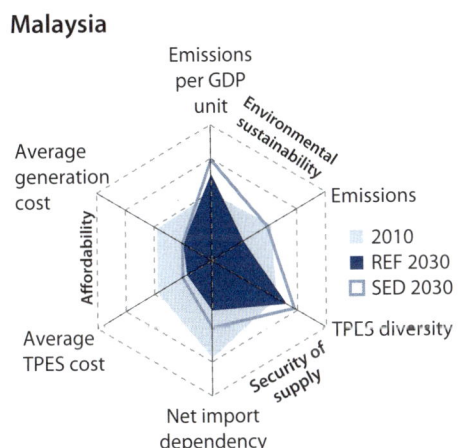

Philippines

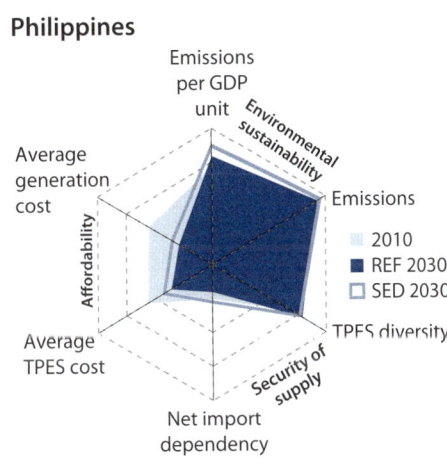

Thailand

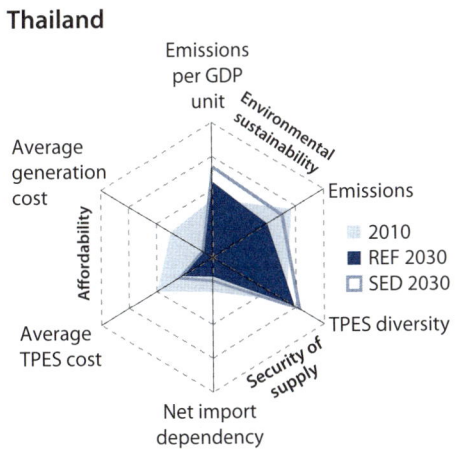

Vietnam

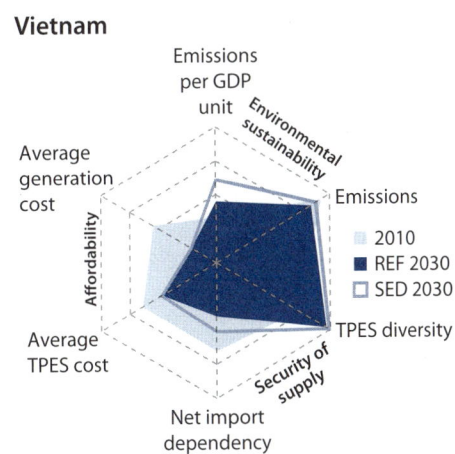

Source: Authors.

In the SED scenario, by 2030, the average electricity costs would increase by 2 percent compared to the REF scenario. For most consumers, the impact of higher unit costs will be offset by rising incomes and more efficient use of electricity. At the national level, countries with diversified fuel mixes will have a declining cost of energy as a percentage of GDP. Moving from the REF to the SED scenario, the average cost of energy as a percent of income would fall by 1.5 percent in the region.

However, low-income consumer groups may face affordability issues. Assuming that income, price, and consumption trends among the poorest households follow the national trends, affordability for these households will improve under the SED scenario relative to REF. Based on the limited household expenditure data available,[43] energy costs as a share of household income are likely to increase only in the Philippines. There, by 2030, energy costs may approach 5 percent of the income of households in the lowest quintile. In areas or communities that experience slower income growth than the national average, more limited access to energy efficiency improvements, or more rapid increases in prices for energy supplies to poorer customers in remote areas, affordability is likely to remain a concern.

In sum, the radar diagrams (figure 4.21) show the changes in indicators of environmental sustainability, energy security, and costs under the SED scenario in each major East Asian economy in 2030, compared to the REF scenario. Higher scores (farther from the center) indicate better performance (lower emission, higher diversity, lower import dependency, and lower cost).

---

43. As noted earlier, a detailed analysis of affordability would require data and projections on the statistical distribution of household incomes and household energy expenditure. These data and type of analysis were not part of this study, nor of the parallel household energy study. However, the latter provided limited data on household incomes and energy expenditure levels.

5

Path to Energy Sustainability: Transformative Policy Tools and Financing Mechanisms

Key messages:

- Shifting to a sustainable energy path requires East Asian governments to take immediate action on major policy and institutional reforms to transform the energy sector toward much higher energy efficiency and more widespread use of low-carbon technologies. To fully realize the huge energy efficiency potentials in the region depends on policy and institutional reforms to overcome market failures and barriers.

- To achieve this sustainable energy path also requires transferring substantial financing and low-carbon technologies from developed countries. A major hurdle is to mobilize financing for the net additional investment of $80 billion per year over the next 2 decades. Approximately $25 billion per year are required as concessional financing to cover the incremental costs and risks of energy efficiency and renewable energy. In addition, substantial grants are needed to build the capacity of local stakeholders and provide technical assistance.

- While many East Asian countries are taking steps to get onto a sustainable energy path, accelerating and scaling up these efforts are needed. The window of opportunity is closing fast. Delaying action would lock the region into a high-carbon infrastructure, requiring future costly retrofitting and premature scrapping of existing energy stocks.

- Policy tools and financing mechanisms exist for such transformations, but they need to be tailored to the maturity and costs of technologies and national context. Only strong political will and unprecedented international cooperation will make them happen.

5.1 Immediate government action is required

To Act Now Will Lock the Region's Energy Infrastructure into a Low-Carbon Future

Over the next decade, new power plants, buildings, roads, and railroads will lock in technology and largely determine emissions through 2050 and beyond. Energy capital stock has a long life. It can take decades to turn over power plants, a century to turn over urban infrastructure (Shalizi and Lecocq 2008) (figure 5.1). Delaying action would substantially increase future mitigation costs—effectively locking the region into carbon-intensive infrastructure for decades to come. Even existing low-cost clean energy technologies will take decades to fully penetrate the energy sector. Given the long lead times required to develop new technology, deploying advanced technologies on a large scale by 2030 requires major policy actions today (World Bank 2009a).

Delaying action would, in addition, lead to costly retrofitting and early retirement of the energy system. Building to current standards and then retrofitting existing capacity, whether power plants or buildings, would be far more costly than building new, efficient, low-carbon infrastructure in the first place. The same is true for the forced early retirement of inefficient energy capital. Energy savings often justify the higher upfront investments in new capital, but they are less likely to cover premature replacement of capital stock. Even a high CO_2 price may be insufficient to change this picture (Philibert 2007).

Figure 5.1 Energy Stocks Have Long Lifetimes and Slow Turnover *(no. of yrs)*

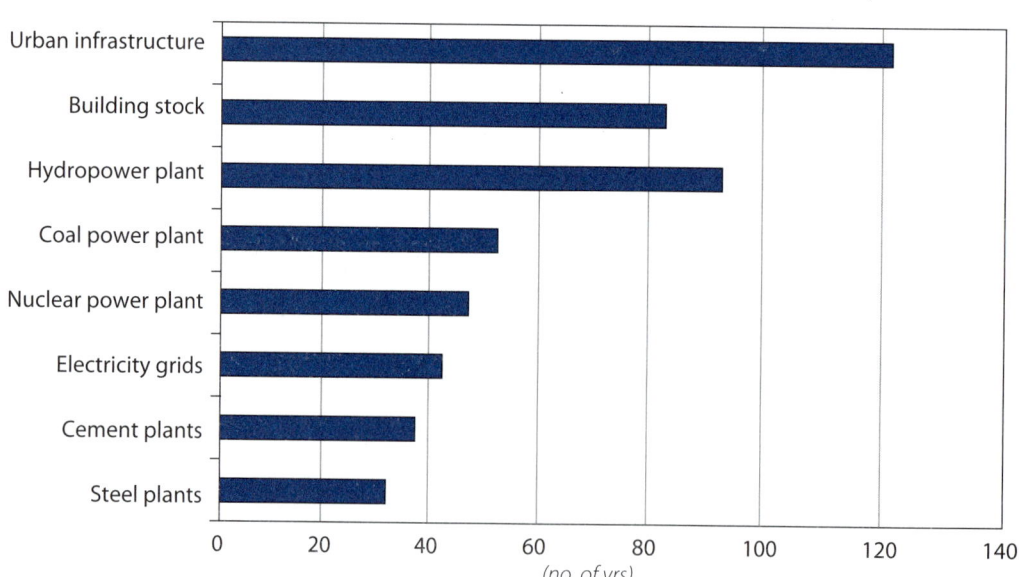

Sources: Authors based on data from 2008b.

Figure 5.2 China Has One of the World's Largest Green Stimulus Plans *(%)*

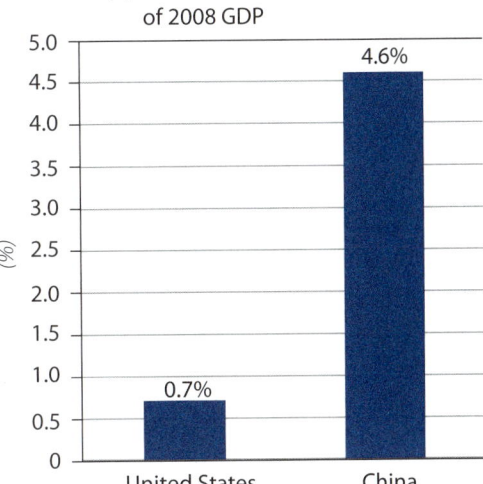

5.2(a) Economic stimulus dollars devoted to green projects

5.2(b) Green stimulus dollars as a ratio of 2008 GDP

Source: HSBC Global Research 2009.

One benefit of the inertia in energy infrastructure is that introducing efficient low-carbon technologies into new infrastructure offers an opportunity to lock in a low-carbon growth path. Over the next decade, developing countries will install at least half of their long-lived energy capital stocks (McKinsey Global Institute 2009a). The scale and rate of urbanization in East Asia present an unrivaled opportunity to make major decisions today about building low-carbon cities. These would require compact urban designs, good public transport, green buildings, and clean vehicles.

Financial Crisis Offers an Opportunity for Efficient and Clean Energy

Governments should not use the current financial crisis as an excuse to delay climate change actions, because the future climate crisis is likely to be more damaging to the world economy. The economic downturn may delay the business-as-usual (BAU) path by a few years, but it is unlikely to fundamentally change that path over the long term.[44] Instead, the downturn offers opportunities for governments to provide green stimulus investment in efficient and clean energy to meet the twin goals of revitalizing economic growth and mitigating climate change (World Bank 2009a) (figure 5.2). In some East Asian countries, fossil fuel subsidies or contingent liabilities on national governments from the energy sector already are a significant percentage of GDP. For these nations, additional "green

---

44. McKinsey Global Institute 2009b; Blanford and others 2009, S82–93.

Table 5.1 Many East Asian Countries Have National Plans and Sustainable Energy Targets

Country	Energy efficiency	Renewable energy	Carbon reduction
China	20% reduction in energy intensity from 2005–10	15% of primary energy by 2020	40%–45% reduction in carbon intensity from 2005–20
Indonesia	30% energy efficiency improvement from BAU by 2025	17% of primary energy from renewable by 2025	26% reduction in carbon emissions from BAU by 2020; 41% with international support
Thailand	1. Energy conservation fund and ESCO fund 2. EGAT demand side management program 3. Appliance standards and labeling	20% of final energy demand from renewable by 2022	Bangkok Metropolitan Administration (BMA) set a target to reduce the city's carbon emissions by 15% per year from 2007–12
The Philippines	Annual energy savings of 2.9 Mtoe from 2005–14[a]	Doubling renewable capacity by 2030	
Vietnam	3%–5% reduction in total energy consumption by 2010, 5%–8% reduction by 2015	3% and 5% of power capacity from new renewable[b] by 2010 and 2020	

Sources: Government of China 2008; Chinese government's communication to UNFCCC 2010; Indonesian government's communication to UNFCCC 2010; Clean Technology Fund Investment Plan for Indonesia; Clean Technology Fund Investment Plan for Thailand; Clean Technology Fund Investment Plan for the Philippines, Department of Energy of the Philippines; and Clean Technology Fund Investment Plan for Vietnam.
Notes:
a. Mtoe = million tons of oil equivalents.
b. This does not include large hydropower plants, which accounted for 26% of existing power capacity.

stimulus" financing will be necessary to shift to clean energy technologies.

In conclusion, East Asian governments need to take immediate action to transform their energy sectors into high energy efficiency systems in which the use of low-carbon technologies is widespread. A paradigm shift is required to leapfrog to new development models and shift lifestyles toward conservation and sustainability.

East Asian Governments Are Taking Action toward Sustainability

Many countries in the region are taking actions, adopting policies, and setting targets for clean energy technologies (table 5.1). Many of these initiatives are driven by domestic development benefits: more energy savings, less local air pollution, greater energy security, more employment in local industry, greater competitiveness from higher productivity, and less traffic congestion. These benefits can justify part of the mitigation cost and increase the appeal of green policies.

Figure 5.3 China Has World's Largest Renewable Energy Capacity *(GW)*

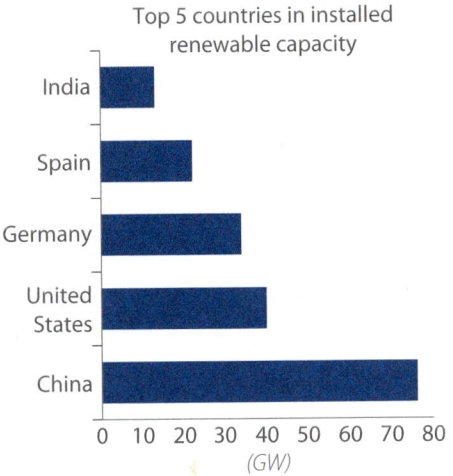

Source: Authors based on data from REN21 2009.

Figure 5.4 Philippines and Indonesia Have World's Second and Third Largest Geothermal Power Capacity *(MW)*

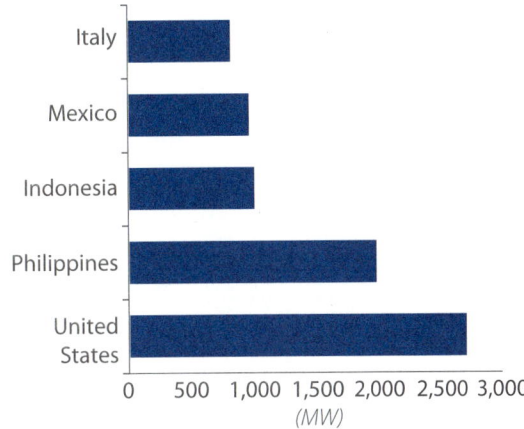

Source: Authors based on data from REN21 2009.

They also can substantially reduce CO_2 emissions. In 2006 the renewable energy industry created 2.3 million jobs worldwide (directly or indirectly), and energy efficiency added 8 million jobs in the United States (EESI 2008).

From 1995–2004, China's energy efficiency efforts reduced its energy intensity by 30 percent—a savings equivalent to Japan's annual total energy use. The Chinese government's target of a 20 percent reduction in energy intensity from 2005 to 2010 was intended to reduce its annual CO_2 emissions by 1.5 billion tons by 2010. This is the most aggressive emissions reduction target in the world—5 times the 300-million-ton reduction of the European Union's Kyoto commitment. From 2006–09, China achieved a 14.4 percent reduction in energy intensity. Furthermore, China has the world's largest renewable energy capacity (figure 5.3). Renewables account for 8 percent of its energy and 17 percent of its electricity, thanks largely to small hydro. China also has targets to increase renewables' share of primary energy to 15 percent by 2020.

The Philippines and Indonesia have the second and third largest geothermal capacity in the world (figure 5.4 and box 5.1). The Philippines also has the third largest biomass power capacity in the world, and plans to double its renewable energy capacity by 2030 (REN21 2008). Indonesia and Thailand target to increase their shares of renewable energy in their total energy to 17 percent by 2025, and to 20 percent by 2022, respectively.

However, a large gap remains between the direction in which the region is headed under current policies and the direction in which

Box 5.1 Geothermal Energy Development in the Philippines and Indonesia

Geothermal power is one of the best options as an indigenous energy source to reduce reliance on energy imports in the Philippines, and diversify energy mix in Indonesia. Geothermal is a baseload generation technology, providing power on a "24/7" basis not subject to the intermittence and variability of most renewable electricity sources. As a result, geothermal can directly displace additional coal-fired power generation. As an indigenous energy source, it also will enhance the country's energy security and serve as a natural hedge against the volatility of fossil-based commodity prices.

The Philippines

The Philippines already has the second largest geothermal capacity in the world, with an installed capacity of 2,000 MW, which provides 17 percent of the country's power generation (box figure). Geothermal power was first produced in the Philippines in 1979 with the development of the Tiwi and Mak-Ban fields in Southern Luzon by Philippine Geothermal, Inc., a subsidiary of Union Oil Company of California. Other geothermal fields in Luzon, Mindanao, Leyte, and Southern Negros subsequently were developed by the Philippine National Oil Company Energy Development Corporation (PNOC-EDC). In the early 1990s, the country experienced a rapid upswing in geothermal power development. This rise was due largely to national build-operate-transfer (BOT) legislation. These laws enabled international power utilities to enter the market and to fund and construct geothermal power plants. This policy enabled an increase in much needed generating capacity without increasing the national debt.

Since then, the power sector has been unbundled to separate power generation, steam field operations, and transmission and distribution. Philippine law envisions all energy-related activities in the hands of private companies. This decision meant that the state had to tender out the National Power Company (NPC) and PNOC geothermal projects and that these companies were not allowed to build new ones. A spot market has been founded. Without priority being given to renewable energy, the risk is that the market will prefer cheaper power from coal over geothermal. Another issue related to the spot market will be the shift

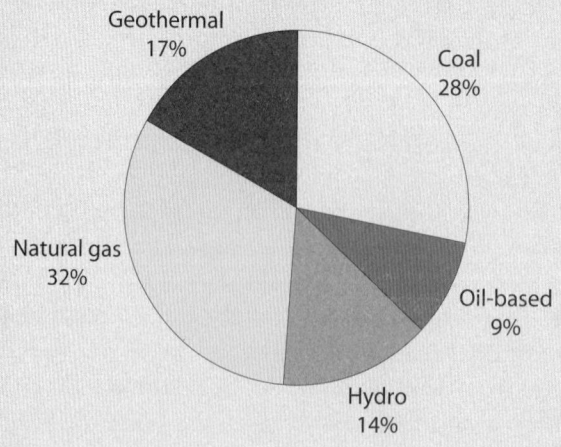

away from long-term PPAs. They were important to obtain project financing. Finally, due to the Asian financial crises a decade ago, funding geothermal projects was put on hold.

In 2008 the Philippines president signed the Renewable Energy Act. In 2009 the Department of Energy issued the Implementing Rules and Regulations. This Act will activate renewable energy development by creating a market conducive to energy generation. In response, there are a long list of projects being developed and a long list of companies that want to enter this promising market.

Indonesia

The Government of Indonesia has developed a blueprint with a target to develop at least 6,000 MW of geothermal power capacity by 2030 from approximately 1,000 MW now. In 2003 Indonesia issued a geothermal law (Law 27/2003), which made geothermal the only renewable energy governed by its own law. The government is in the midst of a long-term reform program in the sector. These reforms include the development of a pricing and incentive policy for geothermal, and the transparent and competitive tendering of new geothermal concessions in line with the Geothermal Law. The government has a two-pronged strategy to develop geothermal energy: (1) improve the investment climate to attract available investments in private concessions, and (2) in the short term, scale up public financing of geothermal fields presently under the control of state-owned enterprises. The Bank Group is assisting the government to scale up geothermal development with a combination of funds from the Global Environment Fund (GEF), World Bank (IBRD), Clean Technology Fund (CTF), International Finance Corporation (IFC), and Carbon Finance.

Source: ESMAP 2009.

the region needs to go to achieve a sustainable energy path. Strong government commitment and political will are essential for such transformations.

5.2 Policy tools and financing mechanisms need to be tailored to the maturity and costs of technologies

Policy tools and financing mechanisms exist for large-scale deployment of energy efficiency and low-carbon technologies, but they need to be tailored to the maturity and costs of technologies and national context (figures 5.5 and 5.6 and table 5.2).

- *Energy efficiency*. In the short term, the largest and cheapest source of emission reductions is increased energy efficiency on both supply and demand sides in power, industry, buildings, and transport. Many energy-efficiency measures are financially viable for investors but have not been fully realized. Realizing these low-cost savings requires pricing reforms, regulations such as efficiency standards and codes, combined with institutional reforms and financing mechanisms to correct market failures and barriers. If the right policy and regulatory frameworks are in place, the bulk of the financing needs in energy efficiency could be met by domestic investment from the public and private sectors. International concessional financing also is required to cover the incremental risks and transaction costs, and to build the capacity of financial institutions and energy service providers.

- *Renewable energy.* In the short to medium term, the second largest source of emission reductions comes from renewable energy for power generation. Many of these technologies are commercially available and can be deployed much more widely under the right policy and regulatory frameworks. Energy storage technology and smart grids can enhance reliability of electric networks when incorporating a large share of variable renewable energy, and distribute generation. Most renewable energy technologies are economically viable but not yet financially viable. Therefore, some form of subsidy to internalize externalities is needed to make them cost competitive with fossil fuels. Adopting these technologies on a larger scale will require incorporation of external costs in fossil-fuel prices through an energy tax or a carbon tax, and financial incentives to adopt low-carbon technologies (World Bank 2009a). While the baseline costs will be borne by domestic financing sources, international concessional financing is needed to cover the incremental costs (costs above fossil fuels).

- *New technologies* (such as Integrated Gasification Combined Cycle, or IGCC; CCS; electric vehicles; energy storage; and smart grids). While commercially available technologies can meet the bulk of the abatements up to 2030, deep emission cuts beyond 2030 will require developing and deploying new technologies at unprecedented scale and speed. Given the long lead time needed for technology development, now is the time for governments to ramp up research, development, and demonstration efforts to accelerate the innovation and deployment of advanced technologies. Historically, innovation and technology breakthroughs have reduced the costs of overcoming formidable technical barriers, given effective and timely policy action—a key challenge facing the world today. The largest barrier is the high incremental costs between these technologies and conventional options, particularly in developing countries. Effective, innovative, fair, and affordable ways are needed to accelerate the transfer of low-carbon technologies and the financing of incremental costs of these technologies to the developing world.

Figure 5.5 Policy Tools Need to Be Tailored to Maturity and Costs of Technologies

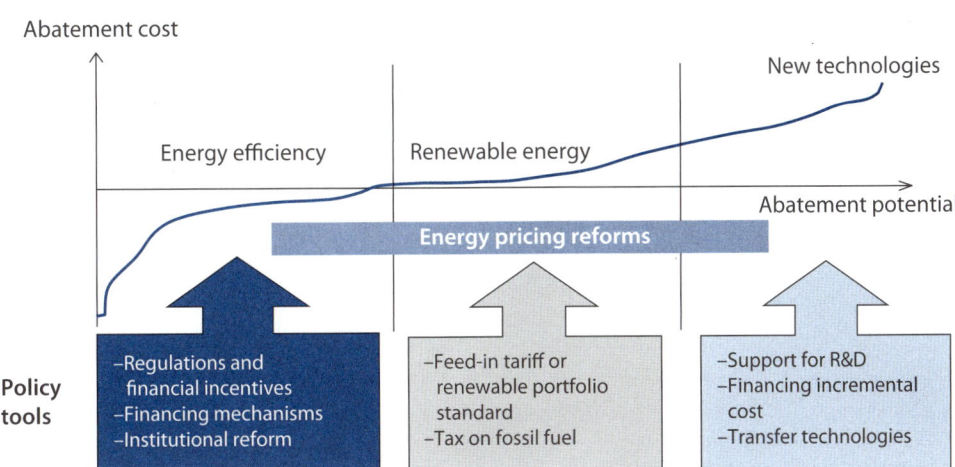

Source: Authors.

5.3 Price should reflect full cost—removing fossil fuel subsidies and internalizing environmental costs

Market-based pricing reforms in the power, coal, oil, and natural gas subsectors, particularly sound price-setting methodologies and tools, are fundamental to an efficient, sustainable, and secure energy sector (Berrah and others 2007). The energy price can be a driving force to discourage consumption, mitigate rebound effects, and

Figure 5.6 Concessional Financing Is Critical

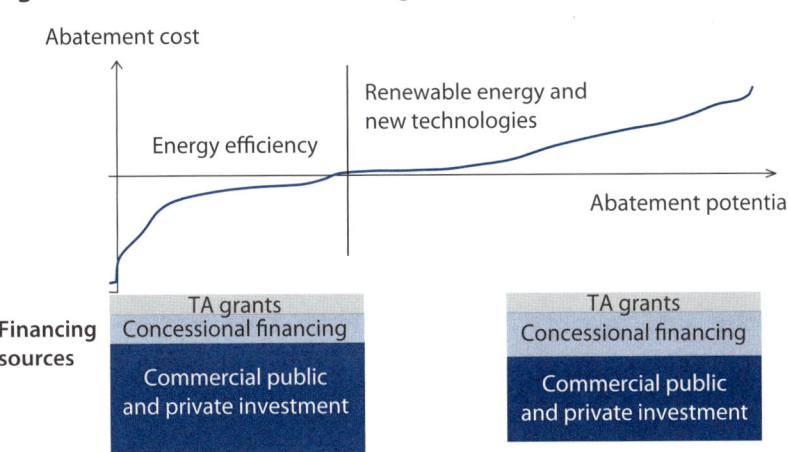

Source: Authors.

Table 5.2 Policy Instruments Need to Be Tailored to Maturity and Costs of Technologies

Maturity level	Description/definition	Issues to address to move to next stage	Policy support
Technically viable (new technology)	The basic science is proven and tested in the lab and/or on a limited scale. Some technical and cost barriers remain.	Development and demonstration to prove operational viability at scale and to minimize costs. Internalize global externalities.	Technology development policies: • Substantial public and private R&D, and large-scale demonstration • Internalize global externalities through carbon tax or cap and trade • Technology transfer.
Commercially available and economically viable (mostly renewable energy)	The technology is available from commercial vendors. Projected costs are well understood. Technology is economically viable justified by country's development benefits. However, it cannot yet compete against fossil fuels without subsidy and/or internalization of local externality.	Lack of level playing field between clean energy and fossil fuels, and internalization of local environmental costs required.	Domestic policies to provide a level playing field: • Remove fossil fuels subsidies and internalize local externalities • Provide financial incentives for clean energy technologies.
Financially viable (mostly energy efficiency)	Technology is financially viable for project investors—cost-competitive with fossil fuels, or with high financial returns and short payback period for demand options.	Market failures and barriers hamper accelerating adoption through the market.	• Regulations, with financial incentives to remove market failures and barriers • Support for delivery mechanisms and financing programs to expand adoption • Consumer education.

Source: World Bank 2009a.
Note: A given technology group can be progressing through different stages at the same time, but in different country settings and at different scales.

encourage clean fuels. Prices should (1) remove fossil fuel subsidies, (2) internalize environmental costs through environmental taxes, and (3) provide incentives to invest end-use energy efficiency.

Removing Fossil Fuel Subsidies

Five East Asian countries—China, Indonesia, Malaysia, Thailand, and Vietnam—are among the top 20 non-OECD countries with the highest energy subsidies—estimated at approximately $70 billion a year, or approximately 2 percent of their GDP in 2007 (IEA 2008a).

The lion's share of the subsidies artificially lowers the prices of fossil fuels. Removal of fossil fuel subsidies could free the fiscal space to pay for the additional financing needs of $80 billion to shift to a sustainable energy path.

In these countries, fossil fuel prices do not reflect international market prices of traded fuels. In some cases, they do not even fully cover the financial costs of production. Subsidies of domestic prices, compared to international prices, have been particularly prevalent in oil products (gasoline, diesel and kerosene) and gas. Some countries such as Indonesia still are subsidizing electricity prices below the costs of production. The subsidies were particularly large during the price spikes of 2008, when most East Asian countries with regulated fuel prices did not raise their domestic regulated prices at the same rate as the spikes.

These subsidies discouraged energy savings and added a significant burden to government's budgets. Most of these subsidies were not targeted at the poor—they applied to all or most consumers of the subsidized fuels, not just those who could least afford them. In 2008 most East Asian countries announced plans to reduce fossil fuel subsidies and raise domestic prices toward international levels. However, actual progress was slow. The situation was largely remedied not through government action, but through sharply falling international fuel prices at the end of 2008.

Falling energy prices provide a unique opportunity to implement programs to eliminate fossil-fuel subsidies in ways that are politically and socially acceptable. Indeed, some countries are now setting prices above the market level to compensate for under-recovery in previous years and to raise taxes.

Removing fossil-fuel subsidies would reduce energy use, encourage clean energy, and lower CO_2 emissions. Ample evidence shows that higher energy prices induce substantially lower demand (World Bank 2008c). This study found that removal of the existing fossil fuel subsidies can contribute a 6 percent emission reduction toward the emission stabilization goal in East Asia. A progressive removal of the subsidies beginning in 2009 would reduce emissions in China by 2 percent, Indonesia by 3 percent, Malaysia by 7 percent, and Vietnam by 10 percent by 2030, as a result of lower energy use and relative price increases for fossil fuels.

However, removing these subsidies is no simple matter—it requires strong political will. Fuel subsidies often are justified as protecting poor people, even though most of the subsidies go to better-off consumers. Effective social protection targeted at low-income groups, in conjunction with the phased removal of fossil-fuel subsidies, can make reform politically viable and socially acceptable. It also is important to increase transparency in the energy sector by requiring service companies to share key information, so that the governments

Box 5.2   Indonesia's Success in Targeting Subsidies to Lessen Impact of Rising Fuel Prices on the Poor

Indonesia has illustrated how to mitigate effect of the rising fuel prices on the poor. Historically, Indonesia's government controlled domestic fuel prices, including gasoline, diesel, and the kerosene used by 90 percent of Indonesian households for cooking. In fact, until 2005, the fuel subsidy, a universal price subsidy, was the centerpiece of Indonesia's social protection scheme. However, the structure of the subsidy was regressive, with the top income decile receiving more than five times what the bottom income decile received (box figure 1). Global fuel prices, which began to sharply increase in 2004 and eventually reached historically high levels, made for a ballooning fiscal burden that became increasingly untenable. In response, the government began to adjust domestic prices, and in some instances, removed the subsidy. Key changes were phased in over 9 months beginning in December 2004. They included a 29 percent fuel price increase in March 2005, the introduction of market prices for industry, and a 114 percent fuel price increase in October 2005 (box figure 2).

It was clear that the proposed price changes, which would free-up budget resources, also would impose a significant burden on the poor. Therefore, in 2005, the fuel price increases were combined with a better targeted cash subsidy that was provided directly to those affected. At the time, it was the largest cash-transfer program in the world, eventually reaching 19.2 million poor households.

The main reasons behind the success of the difficult policy measures that Indonesia implemented include:

- *Careful planning.* The government invested several months in assessing the magnitude of the subsidy reform challenge, including benchmarking the cost of its energy subsidies against those of other countries. It concluded that Indonesia spent 8 to 10 times more on energy subsidies than other countries did.

- *Detailed analysis of options.* The government ensured social protection in implementation strategies by lending targeted support to the middle quintile and the two poorest quintiles.

- *Well-designed "Cash Transfer Program."* The government carefully designed an unconditional cash transfer (UCT) program to make cash transfers to low-income households through the postal system. The program was launched with a widespread communication campaign through broadcast and print media.

During implementation of the program, the Government of Indonesia responded quickly to problems, and addressed them through field assessment, organizing public hearings, establishing a complaint-resolution mechanism, and improving the decentralized distribution of the UCTs.

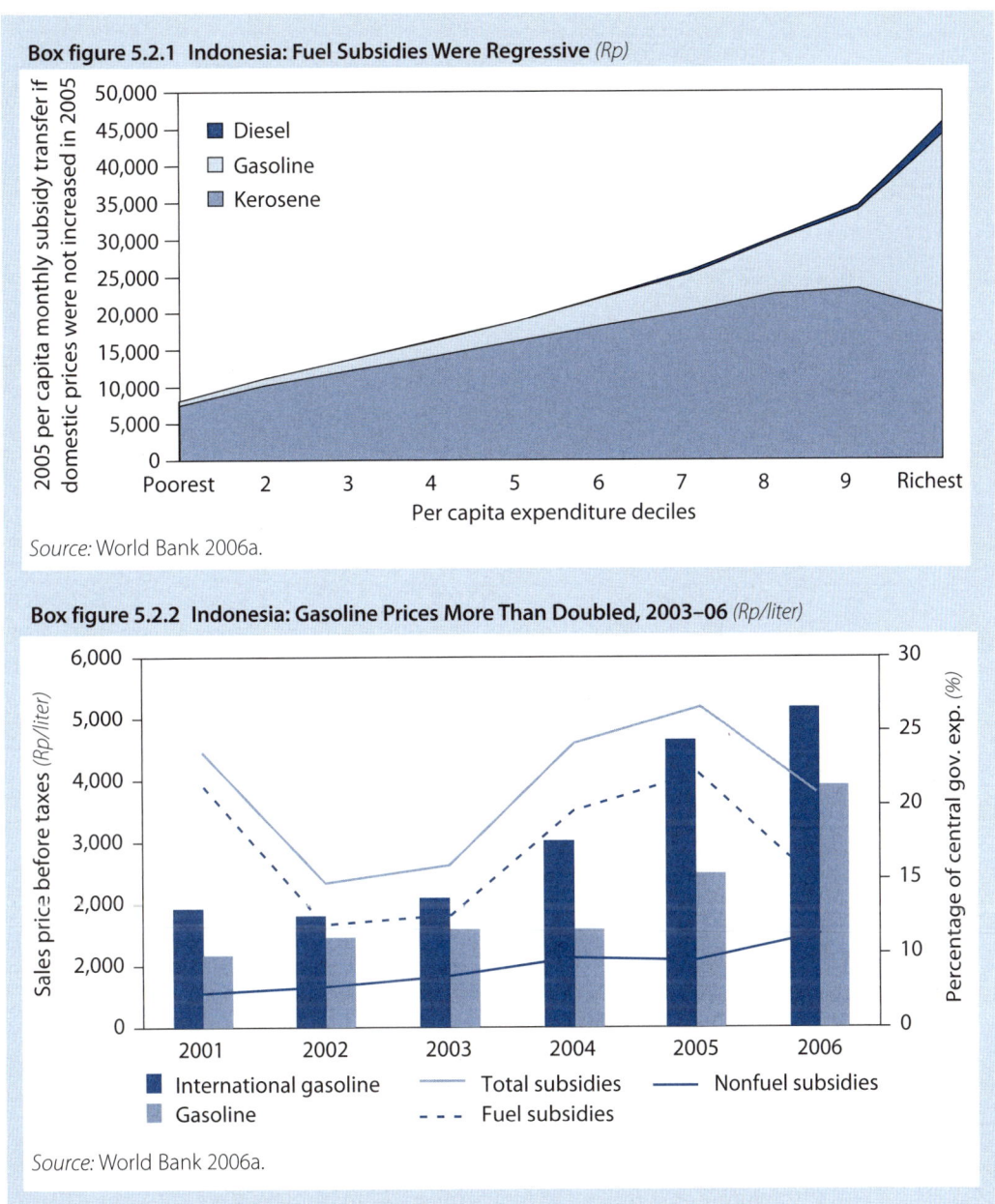

Box figure 5.2.1 Indonesia: Fuel Subsidies Were Regressive *(Rp)*

Source: World Bank 2006a.

Box figure 5.2.2 Indonesia: Gasoline Prices More Than Doubled, 2003–06 *(Rp/liter)*

Source: World Bank 2006a.

and other stakeholders can make better informed decisions and assessments about removing subsidies (World Bank 2009a). Indonesia has successfully targeted subsidies for the poor to mitigate the impacts of rising fuel prices (box 5.2).

Internalizing Environmental Costs

Energy prices should reflect the cost of production. They also should incorporate local and global environmental externalities

> **Box 5.3 China's Coal Tax**
>
> In China, Shanxi Province has implemented a de facto coal tax. It established the Coal Sustainable Development Fund in 2007 and charges all coal produced in Shanxi Province. The surcharge ranges from RMB 5-15/ton (US$0.7–2.2/ton) for steam coal, RMB 10-20/ton ($1.5–3.0/ton) anthracitic coal, and RMB 15–20/ton ($2.2–3.0/ton) for coking coal. In 2008 the total collection was RMB 14.5 billion ($2 billion), which provided an important funding source for energy efficiency and environment protection investments.
>
> _____
>
> *Source:* Authors.

through appropriate use of a fuel tax and/or a carbon tax to address market failures and discourage consumption to mitigate rebound effects (box 5.3). In principle, it should be possible to produce, through the price impacts of environmental taxes, a similar reduction of emissions as achieved through the direct measures of investment in energy efficiency or low-carbon technologies.

Fuel taxes have proved one of the most cost-effective ways to reduce transport energy demand, along with congestion charges and taxes on vehicles based on kilometers travelled, and higher taxes on light trucks and SUVs. For example, if Europe had followed the U.S. policy of low fuel taxes, Europe's fuel consumption would have been twice as large as it is now (Sterner 2007).

Pricing carbon through a carbon tax or cap-and-trade system is fundamental to scale up clean energy technologies and level the playing field with fossil fuels.[45] Carbon pricing provides incentives and reduces risks for private investments and innovations in efficient and clean energy technologies on a large scale (Philibert 2007). This study evaluated the impacts of a carbon tax in East Asia through the PAM. A sufficiently high carbon tax can deliver an emission reduction effect similar to increased energy efficiency measures. For example, a carbon tax starting in 2015 and rising to $100/t CO_2 by 2030 could deliver reductions in emission exceeding those from the energy efficiency measures modeled in the SED scenario.[46]

In the near term, however, developing countries are unlikely to have a carbon emission cap. They also have legitimate concerns over protecting the poor from high energy prices and protecting

_____

45. A carbon tax of $50 a ton of CO_2 translates to a tax on coal-fired power of 4.5 cents per kWh or a tax on petroleum of 45 cents a gallon (12 cents a liter).

46. By comparison, the IEA WEO 2008 study foresees a carbon price of $90/t CO_2 by 2030 in its "550 policy scenario" and $180/t in the "450 policy scenario."

domestic industries from global competitiveness. Therefore, over the near term, a carbon tax is likely to be set at relatively low levels in East Asia and should be combined with other policy instruments to stabilize emissions.

Providing Financial Incentives

Prices also need to provide strong incentives to invest in end-use energy efficiency, such as investment subsidies, soft loans, and tax credits. Low-income consumers are most sensitive to higher upfront costs of efficient products. Financial incentives to offset these upfront costs, such as consumer rebates and energy-efficient mortgages,[47] can change consumer behavior, increase affordability, and overcome barriers to market entry.

Utility-managed energy efficiency and demand-side management programs can produce large energy savings. However, these savings occur only if regulatory frameworks provide strong financial incentives for utilities to save, since utilities make more profit by selling more electricity. A successful example from California is to decouple utility profits from electricity sales. Regulators forecast demand and allow utilities to charge a price that would recoup their costs and earn a fixed return based on that forecast. If demand turns out to be lower than expected, the regulator lets prices rise so that the utility can make the mandated profit. If demand is higher, the regulator cuts prices to return the excess to customers. As a result, California has kept its per capita electricity consumption flat for the past 30 years, substantially below the U.S. national average (World Bank 2009a).

5.4 Energy efficiency policy and financing

Pricing policies alone will not be enough to ensure large-scale development and deployment of energy efficiency and low-carbon technologies. Price addresses only one of many market failures and barriers. As discussed earlier, other barriers, such as a lack of institutional capacity and financing, block the provision of energy-saving services. In transport, buildings, and industry, in which adoption is a function of the preferences of, and requires action by, many decentralized individuals, energy demand is less responsive to price signals. The pricing reforms need to be combined with regulations to correct market failures, remove market barriers, and foster clean technology development. A suite of policy instruments

---

47. An energy-efficient mortgage enables borrowers to qualify for a larger mortgage by including home energy-efficiency measures due to their energy savings.

Box 5.4 Japan's Energy Efficiency Successes

Since the first global oil crisis in 1973, energy conservation has been a cornerstone of Japan's energy policy. Since the early 1970s, driven by energy security concerns, Japan's energy intensity has declined by 40 percent. Japan's conservation policies have combined regulation, supportive measures, corporate energy conservation management, and public and private sector R&D. The Energy Conservation Law focuses on energy efficiency in the industrial, commercial, residential, and transport sectors. Economic incentives include preferential low-interest loans, tax incentives, and subsidies for enterprises that introduce highly efficient systems (World Bank 2008d).

The strict enforcement of the Law—administered by the Ministry of Economy, Trade and Industry (METI), particularly the Energy Efficiency and Conservation Division—has been pivotal to the success of Japan's energy efficiency programs. In 1978 the Energy Conservation Center of Japan (ECCJ) was established. The center receives Japanese government and corporate support to assist with research on and implementation of energy conservation programs, accreditation of energy managers under the Law, and dissemination of energy conservation information.

In the industrial sector, the Energy Conservation Law mandates industries to hire energy managers and submit recommendations for energy efficiency improvements. The voluntary environmental action unites the most energy intensive industries in a pact to achieve energy efficiency goals and deliver results.

In the residential and transport sectors, Japan's Top Runner Program sets fuel efficiency targets for 21 designated machinery and equipment products. It searches for the most efficient model on the market, and that model's specifications become the standard at a specified date in the future or within a certain number of years. The energy-saving labeling system has been introduced to inform consumers of energy efficiency of home appliances and to promote energy-efficient products.

After more than 30 years of continued improvement on energy efficiency and conservation, the success of these efforts pushed Japan against an efficiency ceiling. Thus, the "Energy Conservation Frontrunner Plan" of the "New National Energy Strategy" aims to improve energy consumption efficiency by at least 30 percent by 2030 through establishing a positive cycle of technological innovations and social system reforms, and an energy conservation technology strategy.

Source: Authors.

tailored to the specific institutional and political constraints of each country is required to remove barriers to energy efficiency.

Regulations

Economy-wide energy-intensity targets, appliance standards, building codes, industry performance targets (energy consumption per

Box 5.5 Singapore's Energy Efficiency Strategies

Singapore's Energy Efficiency Program is a key national strategy to mitigate GHG emissions and address climate change that emphasizes sharing knowledge and expertise in energy efficiency. Programs supporting research and development for energy efficiency include the Innovation for Sustainability Fund, Going Upstream, Beyond Test-Bedding, and Demonstration. To reduce the energy use for air-conditioning, the Building and Construction Authority and the National Environment Agency are implementing measures to improve energy efficiency in buildings. For example, under the Building Control Act, air-conditioned buildings must be designed with a high-performance building envelope that meets the prescribed envelope thermal transfer value (ETTV) (50W/sq m). The authority, with the National University, reviewed ETTV standards and explored the possibility of extending ETTV regulations to residential buildings. Study findings were used to create minimum Green Mark standards for new buildings, which came into effect in early 2008.

Source: ADB 2009.

unit of output), and fuel-efficiency standards are among the most cost-effective measures. France and the United Kingdom have gone a step further in instituting energy-efficiency obligations that impose legal energy saving quotas on energy companies. To achieve the obligations, companies can act alone or in partnership to support customer investment in efficiency projects, or buy "White Certificates" from registered vendors. Japan has the one of the lowest energy intensities in the world, and its success illustrates government's long-term commitment to energy efficiency (box 5.4). Japan's energy-efficiency performance standards require utilities to achieve electricity savings equal to a set percentage of their baseline sales or load (WBCSD 2008). Similarly, Singapore also has energy efficiency success stories (box 5.5). Another example is the mandatory phasing out of incandescent lights, as by Australia.

Many East Asian countries have adopted mandatory and voluntary appliance efficiency standards. For example, China, Indonesia, the Philippines, and Thailand have adopted standards for lighting and refrigerators. China and Thailand also have put in place efficiency standards for motors (CLASP 2008). China has set energy efficiency targets for the top 1000 industrial enterprises, which account for 30 percent of China's total energy consumption. Meeting

Figure 5.7 China Has Higher Fuel Economy Standards Than Australia and the United States

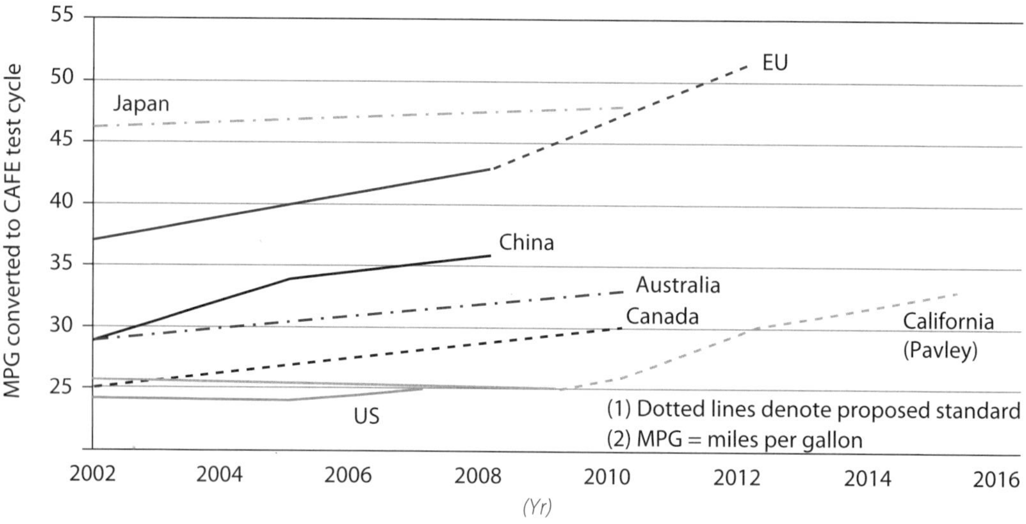

Source: Energy Foundation 2009.

these targets would reduce annual CO_2 emissions by 245 million tons by 2010, higher than California's 2020 emission reduction target of 175-million-ton (Energy Foundation 2009). China's fuel economy standard is higher than that of Australia and the U.S. (figure 5.7). Furthermore, China, Thailand, and Vietnam have energy-efficiency laws. However, as in all contexts, effectiveness depends on enforcement.

Complying with efficiency standards can avoid or postpone adding new power plant capacity and reduce consumer prices. For example, appliance energy efficiency standards in China could save 34 large-scale coal-fired plants in 2020 (figure 5.8). Over the past 30 years, refrigerator efficiency standards in the United States have saved 150 gigawatts in peak power demand—more than the installed capacity of the entire U.S. nuclear power program (Goldstein 2007). Efficiency standards and labeling programs cost approximately 1.5 cents a kilowatt-hour, much cheaper than any electricity supply option (Meyers and others 2005). The average price of refrigerators in the United States has fallen by more than half since the 1970s, even as their efficiency has increased by 75 percent (Goldstein 2007).

In many East Asian countries, weak enforcement of regulations is a main concern. Better enforcement is observed when a small number of stakeholders are involved. For example, refrigerator efficiency standards require monitoring a handful of refrigerator manufacturers. However, when a large number of stakeholders are involved such as with building codes, compliance is a tricky business. Even in California, the building code compliance rate is only approximately

Figure 5.8 Appliance Efficiency Standards Could Avoid 34 Large-Scale Coal-Fired Power Plants in China in 2020

Efficient appliances could save 12% of residential electricity in 2020, or avoid 34 large (1,000 MW) coal-fired power plants

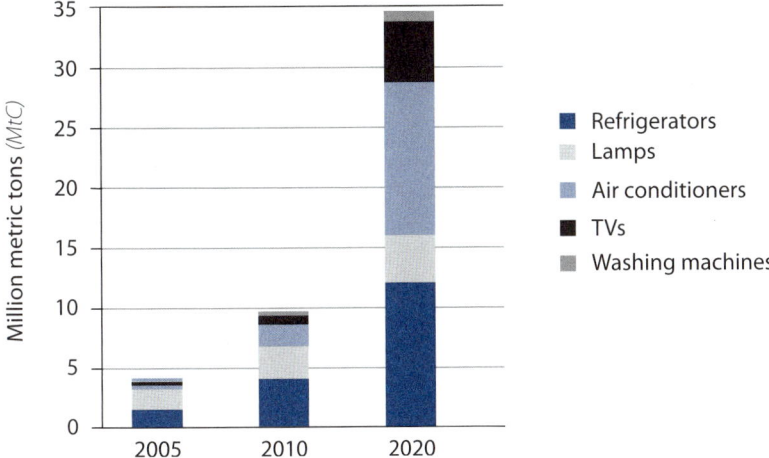

Source: Energy Foundation 2009.

60 percent. Therefore, regulations need to be supplemented with financial incentives for consumers and producers.

Financing Energy Efficiency

Despite the high financial returns and short payback period of many energy efficiency measures, financing an upfront additional invest-ment of $85 billion per year in energy efficiency poses major chal-lenges. These investments tend to be small with high transaction costs. They also are perceived as risky since investors are not sure whether the expected future savings will be realized. Financial institutions usually are not interested and lack expertise in assessing efficiency projects. Lack of domestic capital is rarely a problem. Rather, inadequate organizational and institutional systems for developing projects and accessing funds usually are the barriers.

Therefore, in addition to grants to build capacity of local financial institutions and energy service providers, concessional financing is essential to cover the incremental risks and transaction costs. This study estimated that the required concessional financing could amount to approximately 20 percent of the $85 billion per year, or $17 billion, of the energy efficiency investment needs to mitigate perceived risks, provide financial incentives, and minimize transac-tion costs through risk guarantee, financial intermediary, rebate, soft loans, and carbon financing. The reminder is likely to be met by commercial financing (figure 5.6).

Box 5.6 World Bank Group Experience with Financing Energy Efficiency

The World Bank and the International Finance Corporation (IFC) have financed a series of energy-efficiency financial intermediary and partial risk guarantee projects, mostly in China and Eastern Europe. The IFC pioneered the use of a guarantee mechanism through selected domestic banks with the Hungary Energy Efficiency Guarantee Fund. A Global Environment Facility (GEF) grant of $17 million was used to guarantee $93 million worth of loans for energy efficient investments. No guarantee has been called, giving local banks confidence in and familiarity with energy efficiency lending. A similar approach in China by IFC also succeeded.

The World Bank also has an ongoing energy efficiency financial intermediary program in China, which played a key role in increasing local banks' confidence and capacity in energy-efficiency financing and creating a robust line of business, through learning by doing. The participating domestic banks on-lend IBRD funds ($200M) to large industrial enterprises for energy efficiency investments, and use GEF grant ($14M) to prepare project pipeline and build capacity. In developing a robust energy efficiency pipeline, a carbon financing deal was reached to recover waste heat for Bao Shan Iron and Steel Company, one of the largest in the world.

One of the key lessons of the experience is the importance of technical assistance, particularly at the beginning, to raise awareness of energy efficiency, to provide training and advisory services to the banks in developing financial structures, and to build the capacity of project developers.[1] In Bulgaria the transaction cost of institutional capacity building for both financial institutions and energy service companies—from project conception to financial closure—has been approximately 10 percent of total project costs at the beginning. The annual cost is expected to decline to approximately 5 percent–6 percent later on.

Source: Authors.
Note:
1. Taylor and others 2008.

It is important to distinguish the perceived from the real risks of energy efficiency investments. In the WBG portfolio of energy efficiency risk guarantee programs, for example, GEF grants are used as a risk guarantee fund on the order of 20 percent of the energy efficiency lending portfolio by local banks. However, the real default rate rarely exceeds 1 percent–2 percent. These operations have increased local banks' confidence and capacity in assessing energy efficiency lending and jumpstarted this business line (box 5.6).

The need for risk mitigation and financial incentives varies by sectors. The small and medium-sized enterprises require more risk mitigation, as their perceived risks are higher than those for large-scale enterprises. Individual consumers in the residential sector often demand very short payback times, and financial incentives are sometimes required to offset the higher upfront costs of energy-efficient products. More than one-third of the $85 billion additional financing would be invested in the transport sector, which also requires financial incentives for consumers to shift to more efficient vehicles. The need for concessional financing is expected to decline over time in each country.

In addition to the annual $85 billion investment needs, substantial grants for technical assistance also are critical. Empirical evidence demonstrates that these grants would require approximately 10 percent of the total investment costs to build the capacity of local financial institutions and energy service providers. This need varies greatly by country. Countries with lower institutional energy efficiency capacity and market penetration of energy efficiency require more technical assistance.

Three main financing mechanisms have been developed for energy-efficiency projects: (1) loans and partial loan guarantee schemes operating within commercial banks or as specialized agencies or revolving funds; (2) energy-efficiency and demand-management funds, financed by a surcharge on electricity consumption (system benefit charge) or a government budget, and managed by utilities or dedicated agencies; and (3) third-party financing through Energy Service Companies (ESCOs) (Taylor and others 2008).

Under the first mechanism, lending through local commercial banks offers the best prospect for program sustainability and increasing local banks' confidence in jumpstarting energy-efficiency financing. The WBG has successfully supported financial intermediary and partial-risk-guarantee programs for energy efficiency in China and Eastern European countries (box 5.6). Dedicated revolving funds are another common approach, particularly in countries in which investing in energy efficiency is in the early stages and banks are not ready to provide financing. This approach is transitional, and sustainability is a major issue.

For the second mechanism, a number of East Asian countries including China, Thailand, and Vietnam set up energy efficiency and demand-side management funds. These countries usually face three key issues: (1) where to obtain the funding, (2) who should administer the funds and implement the energy efficiency/DSM programs, and (3) how to most efficiently and effectively use the funds and verify results.

International experience demonstrates that a tariff surcharge from electricity consumers is the most reliable and sustainable source of funding. For example, in the U.S., 30 states have utility energy efficiency/DSM programs financed by the system benefit charge. In 2006 their total combined spending for these EE/DSM programs was $2.2 billion, which ranged from 1 percent–3 percent of each state's utility revenues.

There are several institutional models to manage the energy efficiency/DSM fund: (1) power utilities such as Thailand's EGAT and California utilities; (2) dedicated government agencies such as the New York Energy Research and Development Authority (NYSERDA); (3) energy efficiency utilities such as Efficiency Vermont; and (4) financial institutions in Bulgaria, Hungary, and Romania. The implementing agencies usually are ESCOs, industrial customers, or utilities.

The Standard Offer approach provides an output-based and streamlined mechanism for acquiring demand-side resources (energy efficiency and load management). A Fund Administrator (utility or a public agency) "purchases" energy and/or demand savings using a predetermined and pre-published rate, called the Standard Offer. These rates are based on the value of the energy and demand savings to the utility system, not on the cost of implementation. Any energy user or energy service company (ESCO) that can deliver energy and demand savings is paid the fixed amounts per kWh and kW on completion of the project and certification of the achieved savings.

Finally, carbon financing has had a minimal impact on demand-side energy efficiency. To date, only a 5 percent of carbon market trading has been for demand-side energy efficiency projects. Programmatic and sectoral CDM approaches and innovative ways to use carbon revenues for upfront financing are needed. Such approaches would help overcome the high transaction cost barrier that project-based CDM faces in energy efficiency projects (Sarkar and Singh 2009). Furthermore, the underlying difficulty with energy efficiency financing is that financiers are reluctant to pay cash upfront for uncertain energy savings. While carbon financing does provide additional incentives by enhancing the revenue stream, it does not fundamentally address the upfront financing barrier. The latter can be addressed if carbon revenues from the first batch of energy efficiency projects can be used to finance subsequent energy efficiency projects. The India Chiller Energy Efficiency Project is an example (box 5.7).

Institutional Reforms

An institutional champion, such as a dedicated energy efficiency agency, is essential to coordinate multiple stakeholders and promote and manage energy efficiency programs. In East Asia, this actor would be a government agency with a focus on clean energy or energy efficiency, such as the Department of Alternative Energy Development and Efficiency in Thailand; or an independent corporation or authority, such as the Korea Energy Management Corporation (table 5.3).

To achieve successful results, all of the institutional models require adequate resources, the ability to engage multiple stakeholders, and the ability to maintain independent decisionmaking and credible results monitoring (ESMAP 2008). Ideally, the institutional

> ### Box 5.7 Innovative Carbon Financing for Energy Efficiency Projects in India
>
> The India Chiller Energy Efficiency Project (CEEP) was designed to replace older CFC-based centrifugal chillers by non-CFC-using chillers which are more energy efficient (Hosier and others 2009). Grant funds from both the GEF and the Montreal Protocol Fund are used to provide an upfront subsidy to chiller owners to encourage them to invest in the new equipment. Before incentives are disbursed to chiller owners, they must agree to render any future carbon credits and the related payments to the project. These future carbon credits will be transferred to the project, and the resulting revenues will be managed by the implementing agency as additional incentive payments to replace additional chillers. Of the targeted 370 chillers, 185 are supported by GEF funding and 30 by the funding from the Montreal Fund. Another 155 chillers will be replaced through carbon credits earned from the units replaced by GEF and the Montreal Protocol's Multilateral Fund (MLF) and from chillers replaced in the subsequent years of the project by CDM revenues. With the total investment cost of approximately $90 million, the project will rely on local investment capital to pay the bulk of the replacement costs by the chiller owners themselves or through a loan from a local financial institution. The incentive payments from the GEF, MLF, and CDM, are designed to pay for 20 percent of the replacement costs.
>
> *Source:* Authors.

structure should include a law or other high-level legal instrument that defines the structure, the institutions with clearly defined roles and responsibilities and the resources to implement them, measurable targets, and monitoring these targets.

Market-based mechanisms such as ESCOs are important to the implementation of energy efficiency by delivering technical and financing services. ESCOs serve as project aggregators and provide energy efficiency services and financing to clients. However, most ESCOs have had difficulty in obtaining adequate financing from

Table 5.3 Institutional Structure to Implement Energy Efficiency Programs in East Asian Countries

Country	Agency name	Enabling mechanism	Institutional model
China	National Development and Reform Commission	Energy Conservation Law	Energy efficiency public agency
Japan	Energy Conservation Center	Energy Conservation Law	Energy efficiency public agency
Korea, Rep.	Korea Energy Management Corporation	Rational Energy Utilization Act	State-owned enterprise focused on energy efficiency
Thailand	Department of Alternative Energy Development and Efficiency	Energy Conservation and Promotion Act	Energy efficiency public agency

Source: ESMAP 2008.

commercial banks because of the ESCOs' weak balance sheets and the perceived higher risks of loans dependent on revenues from energy savings. Policies, financing, and technical support from governments and international development banks can strengthen ESCOs and mainstream their business model (World Bank 2009a). In China, for example, after a decade of capacity building supported by the World Bank and the government, the ESCO industry grew from 3 companies in 1997 to more than 400 in 2007 with $1 billion in energy performance contracts (box 5.8).

Public procurement

Mass procurement of EE products can substantially reduce costs, attract larger contracts and bank lending, and lower transaction

Box 5.8 It Took a Decade to Create an ESCO Industry in China

China's first three ESCOs were created in 1997 in Beijing Municipality, Shandong Province, and Liaoning Provinces. Startup was supported with European Commission and Global Environment Facility (GEF) grant assistance. Through the China Energy Conservation Project, a World Bank loan helped finance growth. Beginning in 1998, these three companies successfully pioneered the business adapting the energy performance contracting concept to the Chinese market. A number of other companies began to pick up the successful model.

In November 2003, the Second Energy Conservation Project of the World Bank and Chinese government was launched to help develop a robust Chinese ESCO industry of more companies and an expanding scale. The project includes support for the operation of a loan guarantee program for ESCO projects (started operation in 2003) and the development of the Energy Management Company Association (EMCA)[b] (officially launched in 2004). The $20 million risk guarantee fund with GEF grants has enabled approximately $100 million in energy efficiency lending by local banks, with only 1 percent–2 percent default rate.

One of the most important contributions of the loan guarantee program has been to help many small and medium-sized companies jumpstart energy performance contracting, establish their first credit records, and develop their first borrowing relationship with a bank. The project has helped companies understand lenders' requirements and helped companies properly assess their financial affairs and improve their financial management. A substantial number of China's more famous new ESCOs paid for loan guarantees to obtain initial loans for their business and subsequently grew to become core companies in the new ESCO industry.

During the subsequent three years, China's ESCO industry grew at an astonishingly fast pace, with annual investments in energy-performance-contracted EE projects of approximately US$1 billion in 2007.

Source: World Bank 2008a.

costs. In Vietnam, the bulk procurement of 1 million compact fluorescent lamps (CFLs) substantially reduced the costs of the lamps and improved product quality through technical specifications and warranty. Once installed, they cut peak demand by 30 megawatts.[48] The Vietnam program has a cost-benefit ratio of 99 for the utility and 7 for the consumer, and established a viable CFL market (International Institute for Energy Conservation 2007). Public procurement through government agencies, usually one of the biggest energy consumers in an economy, can reduce costs and demonstrate government's commitment and leadership in energy efficiency. However, mandates, incentives, and procurement and budgeting rules must be in place (ESMAP 2009).

Consumer education

Consumer education can promote lifestyle changes and more informed choices. Examples include EE labeling and increased use of electricity and heat meters, particularly smart meters. Consumer awareness campaigns are most effective in conjunction with regulations and financial incentives.

5.5 Renewable energy policy and financing

Transparent, competitive, and stable pricing through long-term PPAs has been most effective in attracting investors to renewable energy. An enabling legal and regulatory framework can ensure fair and open grid access for independent power producers. Three major mandatory policies for renewable power generation are operating worldwide: feed-in laws, renewable energy portfolio standards (RPS), and competitive tendering (ESMAP 2006).

- *Feed-in laws* require mandatory purchases of renewable energy at a fixed price. Feed-in laws such as those in China, Germany, Spain, and Thailand produce the highest market penetration rates in the shortest period (box 5.9). They are considered most desirable by investors because of their price certainty and administrative simplicity and because they are conducive to create local manufacturing industries. Three methods are commonly used to set prices for feed-in tariffs:

  1. Avoided costs of conventional power generation.
  2. Costs of renewable energy plus reasonable returns.

---

48. Under these bulk procurement programs, each lamp costs approximately $1 instead of $3–5, plus another $1 of transaction costs for distribution, awareness and promotion, monitoring and verification, and testing.

3. Average retail prices. Net metering enables consumers to sell excess electricity generated from their homes or businesses, usually through solar photovoltaics to the grid at retail market prices.

The main risk is in setting prices either too high or low, so feed-in tariffs need periodic adjustment (World Bank 2009a).

- *Renewable energy portfolio standards* (RPS) require utilities in a given region to reach a minimum proportion of power or level of power generation from renewable energy, as in many U.S. states, the United Kingdom, and Indian states. The target is met through utilities' own generation, power purchases from other producers, direct sales from third parties to the utilities' customers, or purchase of tradable renewable energy certificates. The downsides are that (1) this policy lacks price certainty and (2) it tends to favor least-cost technologies[49] and established industry players unless separate technology targets or tenders are in place. Renewable portfolio standards also are more complex to design and administer than feed-in laws (World Bank 2009a).

- *Tendering* involves government-sponsored competitive bidding for renewable energy projects to meet a fixed quantity of renewable power. Long-term contracts are awarded to the lowest priced projects, as in China and Ireland. Tendering is effective at reduced costs. The main risks have been that some bidders underbid and that obligations have not always translated into projects on the ground.

Several financial incentives are available to encourage renewable energy investments: reducing upfront capital costs through subsidies, reducing capital and operating costs through investment or production tax credits, improving revenue streams with carbon credits, and providing financial support through concessional loans and guarantees. Output-based incentives generally are preferable to investment-based incentives for grid-connected renewable energy (ESMAP 2006). Investment incentives per kilowatt of installed capacity do not necessarily provide incentives to generate electricity or maintain plant performance. However, output incentives per kilowatt-hour of power produced promote the desired outcome: generating electricity from renewable energy. Incremental costs of renewable energy over fossil fuels can be passed on to consumers or financed through a levy on electricity called a system benefit charge, a carbon tax on fossil fuel use, or a dedicated fund from government budgets or donors (World Bank 2009a).

---

49. For example, an RPS tends to favor wind energy but discourages solar energy.

Box 5.9 China's Wind Concession and Thailand's Feed-in Tariff

China was one of the first developing countries to pass a renewable energy law. The law set feed-in tariffs for biomass power (box table). However, wind power tariffs were established through a concession process. In 2003 the government introduced wind concessions to ramp up wind power capacity and drive down costs. The winning bids for the initial rounds were below average costs and discouraged both wind developers and domestic manufacturers. Thanks to improvements in the concession scheme and provincial feed-in tariffs, China's wind capacity has doubled every year since 2005. As a result, in 2009 China ranked no. 1 in newly installed wind capacity in the world. In 2009 the government announced feed-in tariffs for wind, differentiated by wind resources (box table). The wind concession schemes provided cost benchmarks for establishing feed-in tariffs.

In Thailand, to promote private investment in renewable energy, the government put in place feed-in tariffs, or a pricing adder on top of the average power purchase price for small power producers (10–90 MW) and very small power producers (less than 10 MW). These measures provide additional tariffs above the base tariff of 2.65 Baht/kWh, or 8.3 cents/kWh (power generated from conventional energy) as an incentive to improve commercial viability of renewable energy projects (box table 5.9.1). As a result, in the last 10 years, Thailand's biomass capacity has grown from nearly 0 to 1.6 GW.

Box table 5.9.1 Wind and Biomass Feed-in Tariff in China and Thailand, 2009

	Feed-in tariff[1] (cents/kWh)			
China				
Wind[1]	Class I	Class II	Class III	Class IV
	7.5	7.9	8.5	9.0
Biomass				10.7
Thailand				
Wind		< 50 kW		> 50 MW
		22.3		19.2
Biomass		< 1 MW		> 1 MW
		9.8		9.2

Sources: Authors. Data based on NDRC 2009 and EPPO 2009.
Notes:
1. Chinese wind feed-in tariff is categorized by regions with different wind resources.

Financing renewable energy. Renewable energy technologies are capital intensive but have negligible fuel costs compared to fossil fuels. Therefore, financing the $25 billion per year for additional investments in renewable energy in East Asia faces a major hurdle. This study estimated that the required concessional financing could amount to approximately one-third of the $25 billion investment, or $8 billion, per year needed for renewable energy. The remainder, comprising $9 billion baseline financing plus $8 billion to be recovered from fuel savings, is likely to be met by commercial financing (figure 5.6). Also critical are additional grants for TA.

In reality, the amount of concessional financing is determined by each country's renewable energy tariff. If the tariff is set high enough to ensure the financial viability of renewable energy projects, there is no need for concessional financing. In such cases, domestic consumers pay the incremental costs between renewable energy and baseline costs. For example, in 2008 Chinese consumers paid an additional 0.2 fen/kWh (or 0.03 cents/kWh) for feed-in tariffs of wind and biomass power generation, or a total of $700 million. However, as the share of renewable energy in the power mix increases, the total incremental cost in the whole power system would rise from minimal to relatively high. This rise could increase financial burdens on domestic consumers, particularly on the poor. Although carbon financing can improve the revenue streams of sustainable energy projects, it is unlikely to meet the bulk of the need. Initial experience with the Clean Development Mechanism (CDM) showed that approximately $1 billion of new CDM projects were registered in the best year (84 percent of these projects were in China) (Capoor and Ambrosi 2009). This amount is equivalent to only 1 percent of the projected net $80 billion in financing needs. Even with CDM reforms, carbon financing is likely to remain a small part of the solution. Therefore, additional concessional financing, such as a scaled-up Clean Technology Fund, is needed to bridge the financing gap.

5.6 New technologies

Proven technologies can meet the bulk of the abatement needed in the short to medium term. However, over the long term (beyond 2030), innovations and new technologies (such as IGCC; CCS; electric vehicles; energy storage; and smart grids) are critical to bend the emission curve. Historically, innovation and technological breakthroughs have reduced the costs of overcoming formidable technical barriers—given effective and timely policy action, a key challenge facing the world today. Acid rain and stratospheric ozone depletion are two of many examples demonstrating that estimates of environmental protection costs based on technology extant before regulation were dramatically overstated.

Developing new technology requires a long lead time. Thus, deploying advanced technologies on a large scale beginning in 2030 requires enhanced and accelerated research, development, and demonstration, coupled with an adequate carbon price, today! Even deploying existing clean energy technologies will take decades to fully penetrate the energy sector.

Developed countries need to take the lead in research and development (R&D) in technological breakthroughs, but developing countries cannot afford to wait. The technological revolution offers an opportunity for developing countries to leapfrog to the next generation of new clean technologies, create local manufacturing industries, and become global technology leaders. These benefits already are the main drivers for aggressive renewable energy development in several developing countries. Traditionally, new technologies are produced first in developed economies, followed by commercial rollouts in developing countries, as with wind energy (Gibbins and Chalmers 2008). However, the rapid advent of R&D in major emerging economies such as Brazil, China, and India has made them global leaders. Brazil is a global leader in the production and consumption of ethanol; China has 3 of the top 10 solar systems manufacturers in the world; and India has top global wind systems manufacturers.

However, the largest barrier to develop and demonstrate new technologies in developing countries is the high incremental costs. Effective, innovative, fair, and affordable methods are needed to accelerate the transfer of low-carbon technologies and the financing of incremental costs of these technologies to developing countries.

Concessional financing to cover the incremental costs is required to make demonstration projects a reality on a large scale in developing countries. For example, CCS technology requires substantially higher capital investment than conventional coal-fired power plants. The near-term priority for CCS should be spurring large-scale demonstration projects to reduce costs and improve reliability. The G-8 has pledged to launch 20 large-scale CCS demonstration plants by 2020, which will require capital investments of $30 billion–$50 billion (IEA 2008f). China has launched a number of IGCC and CCS demonstration projects. Some suggest that a global CCS Deployment Fund should be established to pay for the full incremental costs of CCS, funded by a levy on coal-fired power generation from developed countries (Diringer and others 2008).

Developing countries face additional barriers compared to industrialized countries. These barriers include substantially lower public energy research and development (R&D) funding; intellectual property rights (IPR) concerns; generally lower technical and regulatory capacity; limited policy support; a consumer base with less demonstrated interest and ability to pay premium prices for "green" products; limited access to finance for new technology ventures; and higher costs of doing business (especially for new entrants) (World

Box 5.10 China Produces Supercritical Power Plants at One-Third to One-Half of International Prices

China has made impressive progress in acquiring supercritical and ultrasuper-critical (USC) technology and building the local manufacturing capacity needed to produce more than 100 GW per year. The country started gaining the techno-logical know-how in 1992 by building the first plant utilizing Western technology in Shidongkou, Shanghai, and the second plant, also utilizing Western technol-ogy and World Bank financing, in 2004 in Waigaoqiao, Shanghai.

China then proceeded to acquire supercritical and USC technology from OECD countries, first by purchasing power plants and then by obtaining licens-ing agreements and joint ventures. As a result, all the recently constructed power plants have been built locally at very competitive prices ($500–$600/kW), approximately the same cost as subcritical (World Bank 2009g). China's ability to construct advanced plants at little or no additional cost compared to tradi-tional plants plays a significant role in enabling the quick spread of technology. Cost reductions were achieved by localizing the manufacturing process, includ-ing competition among several suppliers and power-generating companies. Today, China has the world's highest installed capacity of supercritical and ultra-supercritical plants, which are installed in 75 percent–80 percent of new coal-fired plants (Mao 2009). China's 1,000 MW design reflects state-of-the-art technological know-how by global standards and is similar to the most efficient plants being built in OECD countries.

Source: Authors.

Bank 2008e). These factors inhibit both advancing precommercial technologies and technology transfer (TT) to developing countries.

In the meantime, TT to developing countries also can have posi-tive impacts on global cost reduction. For example, the lowest cost manufacturers of solar cells and efficient lighting are in developing countries. China produces supercritical power plants at one-third or one-half of the international prices (box 5.10).

6

World Bank's Role: Support Shift toward Sustainability

> **Key messages:** The World Bank Group is committed to scale up investment and analytical and advisory activities (AAA) in sustainable energy to support the East Asian governments. The Bank Group needs to increase its efforts and focus its future East Asia energy business on energy efficiency, renewable energy, and new technologies. The WBG is well positioned to provide policy advice, facilitate knowledge sharing, and catalyze additional financing to help the governments in the region shift to a sustainable energy path. Better integration of new and existing financing sources (IBRD, IDA, GEF, CTF, and carbon financing) can increase the magnitude and speed of sustainable energy development. Furthermore, regional programs can better facilitate knowledge sharing across countries to disseminate the region's successful experience and introduce international best practice. Finally, close coordination with other international financial institutions (IFIs) and development partners can maximize the effectiveness of policy advice and increase the concessional financing needed to make this shift.

The key findings of this study confirm the World Bank Group's support to the region's governments and the focus of the energy program on energy efficiency and renewable energy, which accounted for 40 percent of EAP's energy portfolio over the past decade. However, to put the energy sector on a sustainable path, a substantial scale-up in both lending and analytical and advisory activities (AAA) in sustainable energy will be required across the WBG.

The future WBG operational energy strategy in EAP will focus on four main areas: (1) energy efficiency; (2) renewable energy; (3) new technologies for sustainable energy, and (4) energy access expansion. These 4 areas are consistent with the 2 objectives of the Bank's proposed corporate Energy Strategy: (1) improving access and reliability of energy supply and (2) facilitating the shift to a more environmentally sustainable energy development path.

The EAP infrastructure sectors can achieve considerable synergies through better cooperation, for example, on building low-carbon cities with an integrated approach: transforming buildings and vehicles (including electric vehicles) to be more efficient, transforming mobility toward mass transit, and transforming urban planning (box 6.1).

The WBG is well positioned to provide policy advice, facilitate knowledge sharing, and catalyze financing to help the governments in the region shift toward a sustainable energy path. To achieve the SED scenario will require policy and institutional reforms and concessional financing. The Bank will step up policy advice, institutional strengthening, and knowledge sharing in sustainable energy. Finally, close coordination with other IFIs and development partners to maximize the effectiveness of policy advice and provide catalytic financing to narrow the gap of the additional concessional financing needs to make this shift.

Given the large financing requirements, additional concessional financing is required to help East Asian countries shift to a sustainable energy path. This study estimated that $110 billion per year

Box 6.1 Ecological Cities as Economic Cities

Eco^2 Cities is a recently launched multisector World Bank initiative that bridges the urban, transport, energy, water, and environment sectors of the Bank. The objective is to enable cities to harness the many benefits of ecological and economic sustainability. The Bank's EAP region, in which this initiative was developed, is taking a strategic approach toward promoting this initiative in China, Indonesia, the Philippines, and Vietnam.

The Eco^2 Cities initiative provides cities with an analytical and operational framework that offers strategic guidance on sustainable and integrated urban development. The initiative also introduces powerful and practical methods and tools that can further strengthen and substantiate good decisionmaking—enabling city leadership to get the most from policies and investments in sustainable urban development.

The Eco^2 framework is flexible and easily customized to the context and priorities of each city. The application of the framework will help each city to chart its own unique action plan, or "Sustainable City Pathway." This pathway will comprise a coordinated and sequential program of:

- Taking important reform, institutional, and policy measures
- Investing in specific "catalyst projects"
- Building capacity of institutions and professional staff.

Many of the actions, reforms, and projects undertaken by a city using this platform can be linked to the Bank's ability to access new funding for sustainability and climate change: (a) GEF grants; (b) Clean Technology Funds (concessional financing); and (c) carbon finance. Together, they enable transformation and innovation in sustainability.

Source: Authors based on World Bank 2009h.

Figure 6.1 Weaving Together Financing Instruments Can Have a Bigger Impact

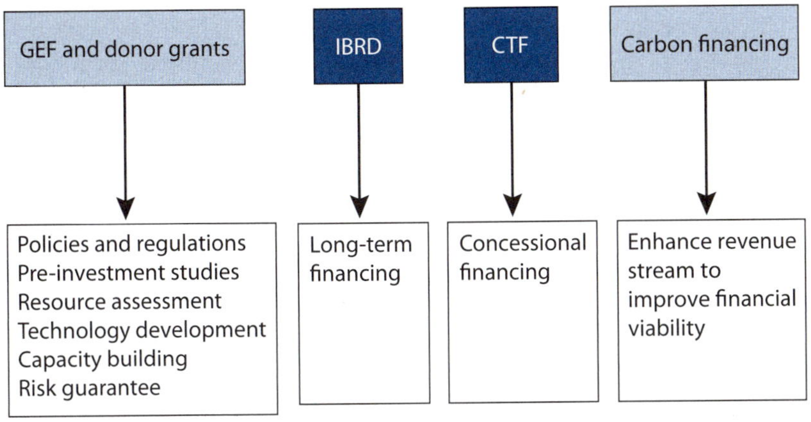

Source: Authors.

additional gross investments are needed in EE and renewable energy for East Asia to shift from the REF to the SED scenario over the next 2 decades. This amount comprises $67 billion in EE and $20 billion in RE in China, $5 billion in EE and $1.5 billion in RE in Indonesia, $4 billion in EE in Thailand, and $2 billion each in EE in the Philippines and Vietnam.

While the public funds from the Bank Group and governments are important, they will be a fraction of the capital costs needed. The Bank will assist the governments to put in place an enabling environment and will work with IFC to leverage private sector investments.

Better integration of new and existing financing sources (IBRD, IDA, GEF, CTF, and carbon financing) can scale up and accelerate the shift to a sustainable energy path (figure 6.1). Grants from GEF and other donors can be used to set up an enabling environment, build capacity, and share transaction costs and risks associated with EE and RE. IBRD/IDA funds are used to provide long-term financing for the capital costs of low-carbon projects. CTF funds provide concessional financing to buy down the incremental costs and risks of low-carbon technologies. Carbon financing adds an additional revenue stream to improve the financial viability of sustainable energy projects.

The regional or subregional programs have the advantages of facilitating knowledge sharing across countries and promoting regional trade. Such programs can send the right signals that the Bank is committed to support sustainable energy in the region. The programs also can help create a large regional clean energy market attractive to private investors and entrepreneurs. The Bank can facilitate dissemination of the region's successful experience and introduction of international best practices. In addition, regional hydropower trade could provide the least-cost energy supply with zero carbon emissions to countries in Southeast Asia.

The immediate follow-up to this study will be country-level policy dialogues and AAA to determine the strategic direction and identify the opportunities in each country program. Given its sheer size and the government's commitment, China leads the WBG's clean energy portfolio of EE and RE. In other East Asian countries, country-specific AAA can help pinpoint the EE opportunities and economically viable potentials of each RE resource, identify barriers, and propose interventions.

Energy Efficiency

All East Asian countries have significant EE opportunities in both the demand side (in the industrial, building, and transport sectors); and supply side (in the power sector such as rehabilitation of coal-fired power plants, fuel switching from coal to gas, and reduction of transmission and distribution losses). While supply-side EE has been mainstreamed in Bank operations, demand-side EE should be strengthened and scaled up in the future, as its potentials are much larger (IEA 2008b).

The WBG can help East Asian countries develop and implement policy and institutional reforms, financing mechanisms, and market-based delivery mechanisms to scale up EE. A decade of WBG EE experience demonstrates that EE efforts must shift from developing technologies to delivering services and savings. In addition, both regulations and financial incentives are required to transform the EE market. The appropriate balance between the two varies from country to country. The major constraints to EE are inherently institutional. Scaling up EE requires a successful institutional framework, technical and management capacity, and strong coordination and cooperation among government at every level (World Bank 2009b).

Financial intermediary and risk mitigation programs have proven successful to mainstream EE financing within the domestic banking sector. In East Asia, local financial institutions have had little experience in financing EE so are unlikely to enter this line of business without external support. The WBG can play an important role in helping domestic financial institutions to increase their confidence and capacity in assessing EE projects by providing risk guarantee, financial intermediary, and capacity building. China's successful experience could be replicated in other East Asian countries. One of the key lessons learned is the importance of TA, particularly at the beginning, to provide training and advisory services to local banks in developing financial structures, and to build the capacity of project developers (Taylor and others 2008).

Initially, EE investments can focus on sectors with high energy savings such as large industrial customers and public/commercial buildings. In most East Asian countries, industry dominates energy consumption and represents the largest potential energy savings. For example, China produces half of the world's cement. The application of the best available technologies in the cement industry could reduce China's CO_2 emissions by 250 Mt—5 percent of the country's total emissions (IEA 2008e). In Vietnam, over the past decade, industry has increased its share of final energy use from 30 percent to 46 percent. Furthermore, the new industrial capacity built over the next seven years will produce more than all of Vietnam's industry today (World Bank 2009f). The Bank's EE programs can focus on energy-intensive equipment (for example, boilers and motors) and/or energy-intensive subsectors (for example, iron and steel, cement, and chemicals) through TA on efficiency standards, sector performance targets, and industry energy management standards, combined with investment in EE implementation.

Buildings represent the second largest energy saving potentials in many East Asian countries. For example, half of China's 2030 urban building stock has yet to be built. Existing EE technologies can cost-effectively save up to 40 percent of energy use in new buildings (Brown and others 2005). The Bank can target commercial and public buildings through TA on policy interventions, business models, and zero-emission building technologies, combined with investments.

Renewable Energy

Scaling up grid-connected RE requires enabling legal, policy, and regulatory frameworks and long-term financing. The number one barrier to scale up renewable energy technologies is that their financial costs are higher than those of fossil fuel power generation. Lessons learned from a decade of WBG experience in grid-connected RE demonstrate three key successful factors: sufficient and stable pricing through long-term power purchase agreements (PPAs), mandatory purchase from the utilities, and passing through incremental costs to consumers (World Bank 2006b).

The WBG can assist East Asian countries to develop and implement such legal, policy, and regulatory frameworks as feed-in tariffs to compensate utilities for the incremental costs of renewable energy until renewable energy costs fall to fossil fuel levels. In addition, the Bank can provide grants to fund pre-investment activities, resource assessment, and capacity building. Even with such a policy framework in place, long-term financing is needed for renewable energy projects due to their capital-intensive nature. It will be important for the Bank to assist governments in inducing local, regional, and international financiers to enter this line of business.

Box 6.2   China Renewable Energy Scale-up Program (CRESP)

The most important factors needed to scale up renewable energy market penetration are to (1) guarantee mandatory grid access, (2) set sufficient tariff levels, and (3) clarify rules to pass through incremental costs for RE. Under the Bank's first RE project, the China Renewable Energy Development Project, the Inner Mongolia wind farm of 100 MW components was cancelled because no agreements on how to share the incremental costs (by nationwide or provincial grids) had been reached. Learning from this experience, under the second RE project, the China Renewable Energy Scale-up Program, or CRESP, through a GEF grant ($40 million) and ASTAE grant, the Bank has been conducting an active policy dialogue with the Chinese government. The Bank introduced international best practices of RE-mandated market policies and assisted the Chinese government in developing RE policy frameworks. In 2006 China became one of the first developing countries to pass a Renewable Energy Law, which requires mandatory purchase of RE by the grids and enables the incremental costs to be shared nationwide.

The policy dialogues were complemented by an IBRD investment ($173 million) in 2 x 100 MW wind farms, a 25 MW biomass power plant, and small hydro projects. This was one of the first such large-scale wind and biomass power plants in China. CRESP introduced best available international technologies and practice. The Inner Mongolia wind farm is not financially viable at the 38 fen/kWh power tariff set by the government. Bank carbon financing played a key role in improving the wind farm's financial viability. Therefore, integrating GEF, IBRD, and carbon financing has made a bigger, more effective impact on RE development in China than would have the same investments made individually.

In addition, CRESP provided cost-shared R&D to domestic wind manufacturers to support joint design with international design institutes to transfer international wind turbine technologies to China. The favorable tariff policies, together with the technology push through cost-shared R&D and the government's requirement of 70 percent local content, have substantially improved the capacity of domestic wind manufacturing industry.

Source: Authors.

Scaling up RE operations also requires a programmatic approach. The China Renewable Energy Scale-up Program (CRESP) is a good example. It integrates policy dialogues, institutional strengthening, project financing, and carbon financing (box 6.2). Another example is the financial intermediary approach used in the Vietnam Renewable Energy Development Project (box 6.3).

As the region ramps up its reliance on RE, increasingly, it will have to address the intermittent nature of power generation from some RE sources such as solar and wind. One approach to increasing reliability is to develop a RE portfolio with complementary resources, such as wind and hydro. Another approach is to mandate the grid companies to purchase RE and provide incentives to compensate their additional costs of addressing intermittency issues.

Box 6.3 Financing Small Renewable Energy Projects in Vietnam

In Vietnam, power generating capacity is forecast to nearly quadruple from approximately 16 GW in 2008 to 60 GW in 2020. The Government of Vietnam recognizes that grid-connected renewable energy projects, which have an estimated potential for small hydro-power alone of 2,900 MW, can complement the generation from large projects. To facilitate RE development, in 2001 the government adopted the Renewable Energy Action Plan and subsequently developed the necessary policy and regulatory frameworks for small RE projects.

Under this framework, for all projects not exceeding 30 MW, a standardized "no negotiations" power purchase agreement (PPA) is now mandatory. The power utility will buy RE power from independent developers on the basis of an "avoided-cost" tariff.

In 2009 the World Bank Board approved the Vietnam Renewable Energy Development Project, which will finance the development of small hydropower plants and biomass, wind, and solar energy projects. This project will provide a refinancing facility to three participating commercial banks for loans to eligible small renewable-based projects. The facility would refinance up to 80 percent of the loans made by these banks to eligible projects. Since the participating bank will take the full credit risk, that bank will appraise the project for both eligibility for refinancing and creditworthiness.

Source: Authors.

Furthermore, in places in which RE resources are located far from load centers, long-range transmission networks are needed. The Bank will assist the East Asian region to develop and implement the schemes and mechanisms needed for utilities to deliver reliable power with intermittent RE generation.

Finally, intraregional trade of hydropower can provide the least-cost energy supply with zero carbon emissions. For example, hydropower generated in Laos can be sold to Thailand (which has limited RE resources) and to Cambodia (which relies heavily on expensive diesel for power generation) (box 6.4).

New Technologies

The WBG can provide cost-shared R&D and facilitate technology transfer (TT) to help East Asian countries leapfrog to the next generation of clean energy technologies. The Bank can tap GEF and other grants to assist in purchasing licenses and to provide cost-shared grants to support applied R&D funding toward promising clean technologies and/or adoption of international quality standards. The China projects have boosted China's solar and wind manufacturing industries (box 6.2). In addition, a guaranteed market approach (an incentive scheme to provide a large guaranteed

Box 6.4 Nam Theun 2: Public-Private Development of Large Hydropower Resources

The 1,070 MW, US$1.45 billion Nam Theun 2 Hydroelectric Project (NT2) in Lao PDR started commercial operations in early 2010. The Lao PDR government is a 25 percent stakeholder in the private sector project, which will be owned and operated by the Nam Theun 2 Power Company Ltd under a 25-year Concession Agreement. NT2 will provide sustainable clean energy primarily for export to Thailand, but also will contribute approximately 75 MW annually to domestic supply. The NT2 revenues, generated through environmentally and socially sustainable development of NT2's hydropower potential, will finance priority poverty reduction and environmental management programs to support the country's goal to graduate to middle-income country (MIC) status by 2020.

Through NT2, the public-private partnership (PPP) model has created opportunities for increased foreign direct investment (FDI) in the country as the private sector and financial institutions have gained familiarity with the Lao context. Lao PDR's ample supply of water, together with its central location in the dynamic Greater Mekong Sub-Region, places the country well to help meet the growing power needs of its neighbors. A large pipeline of hydropower projects under consideration or development is emerging.

Source: Authors.

market to companies that develop and diffuse breakthrough technologies) could substantially reduce the costs of technology development through economies of scale.

Few existing financing sources pay for the high incremental costs of new technologies; hence, additional financing mechanisms are needed. For example, CCS is critical for China's coal-based economy under a carbon constrained future—but is not eligible for CTF.

Cross-Cutting: Policy Advice, Institutional Strengthening, and a Sustainable Energy Knowledge Hub

To accelerate the shift to a sustainable energy path, the Bank needs to ramp up policy advice, institutional strengthening, and knowledge sharing in the areas outlined above. The Bank's value-added in China and EAP5 countries lies in its advisory services and new ideas from international experience and best practices. The Bank needs to increase its efforts to advise its East Asian clients on how to set up an enabling environment conducive to scale up EE and RE, leveraging significant private sector investment and increasing market penetration. Strengthening institutions also is important, particularly innovative institutional models to manage and implement EE. For example, advisory services and knowledge sharing are critical to jumpstart EE lending by domestic financial institutions.

In this regard, this study proposes a World Bank East Asia Sustainable Energy Forum (EASEF). EASEF would engage countries on clean energy policy advice, regional energy cooperation and investments, knowledge sharing, capacity building, and technology development. The forum would promote the sustainability agenda, support harmonization of regional energy markets, provide advisory services, disseminate cutting-edge knowledge, and facilitate South-South cooperation on sustainable energy and climate change.

To raise the profile of sustainability and gain political support, this study also proposes that the Bank's EAP Region call for a high-level Regional Summit. With participation from ministers in EAP countries, this summit would put EE, RE, new technologies, and climate change at the top of the agenda and seek support to establish EASEF.

In conclusion, it is within East Asia's reach to stabilize CO_2 emissions by 2025, improve the local environment, and enhance energy security without compromising economic growth—all through energy efficiency and low-carbon technologies. To achieve these requires immediate action from the governments to undertake major domestic policy and institutional reforms to transform their energy sectors to much higher energy efficiency and more widespread use of low-carbon technologies. Developed countries also need to transfer substantial financing and low-carbon technologies. The technical and economic means exist for these transformations, but only strong political will and unprecedented global cooperation will make them happen. The World Bank Group is committed to scale up policy advice, knowledge sharing, and financing of sustainable energy to help the region's governments make such a shift. The window of opportunity is closing fast—the time to act is now.

Urban Transport Study

The urban transport study conducted in parallel to this East Asia energy report developed a model to determine the fuel consumption and emissions from on-road transport in selected East Asian cities. The transport study examined the potential to reduce transport fuel consumption and emissions through fuel economy standards, urban planning, public transport infrastructure, and pricing policies.

On-road transport is a significant consumer of energy in the urban environment. Of all sectors, on-road transport is the most closely linked to petroleum product consumption. Over the coming years, this sector can be expected to exhibit significant growth in energy requirements in most East Asian countries, in which rising household incomes and urbanization are fuelling private vehicle ownership and use.

The study evaluated the impact of urban transport master plans in 9 cities in the region principally from 2007–20. The 9 cities were Bangkok, Beijing, Chengdu, Guangzhou, Hanoi, Ho Chi Minh City (HCMC) (2 master transport plans), Jakarta, Manila, and Ulaanbaatar.[50] In each city, different levels of action and investment were considered across a range of scenarios. The first was a "Do-Minimal" scenario, in which no significant additional actions were taken to promote public transport other than the capital investment programs currently underway in each city. At the opposite end of the spectrum, a "Do Maximum" scenario was based on the complete implementation of the master plan plus complementary measures to further integrate public transport (focusing on coordination and minimizing the time and cost penalty of transferring between modes) and discourage private car use.

The transport study was based on pre-existing master plans. Thus, it could be concluded that each one had been designed to

---

50. The dates of the studies were dependent on pre-existing models and the work performed by each city. Beijing and Guangzhou had base years of 2005; Beijing and Bangkok had master plan years of 2021; and HCMC had a master plan year of 2025.

Figure A1.1 Growth in Active Vehicle Population by Vehicle Type between Base Year and Year of Master Plan: "Do Minimal" Scenario *(%)*

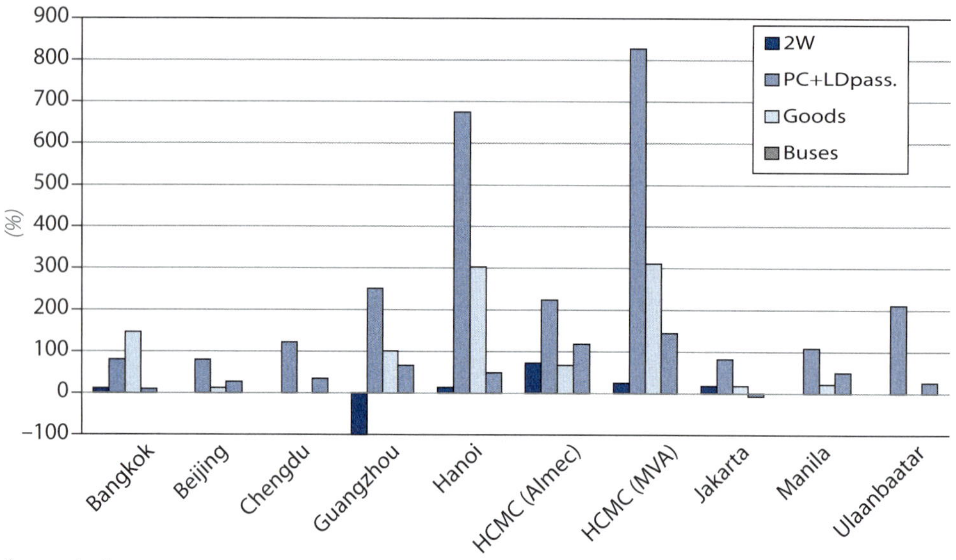

Source: Authors.

resolve specific real-life transport issues within each city and that none of these master plans had been developed to reduce fuel consumption. The cities were chosen as representative of urban transport in the East Asia region. Accordingly, their results may be extrapolated to the overall urban environment experienced in this region. Since the urban transport master plans focus on the movement of people, in most cases, freight may be incompletely modeled.

Vehicle Population

On average, the expected growth in the vehicle population in each city between the base year and the year of the master plan is 72 percent (figure A1.1).[51] Typically, the higher growth occurs in the smaller cities, and the vehicle segments with greatest growth are the passenger car and light-duty passenger vehicles.[52] This projected increase in the total vehicle population varies little across scenarios. Even though improved public transport incentivizes a modal shift from private vehicles, the shift is reflected in reduced vehicle usage, not in a significant change in the level of ownership.

51. A forecast of a 72% increase indicates that the expected active on-road fleet in the year of the master plan is likely to be 172% of that in use in 2007.
52. The light-duty passenger vehicle category includes Sports Utility Vehicles (SUVs), Multi-person vehicles (MPVs), and Asian utility vehicles(AUVs)/Multiple use vehicles (MUVs).

Figure A1.2 Vehicle Use Relative to 2007 Base Year *(vehicle-km travelled)*

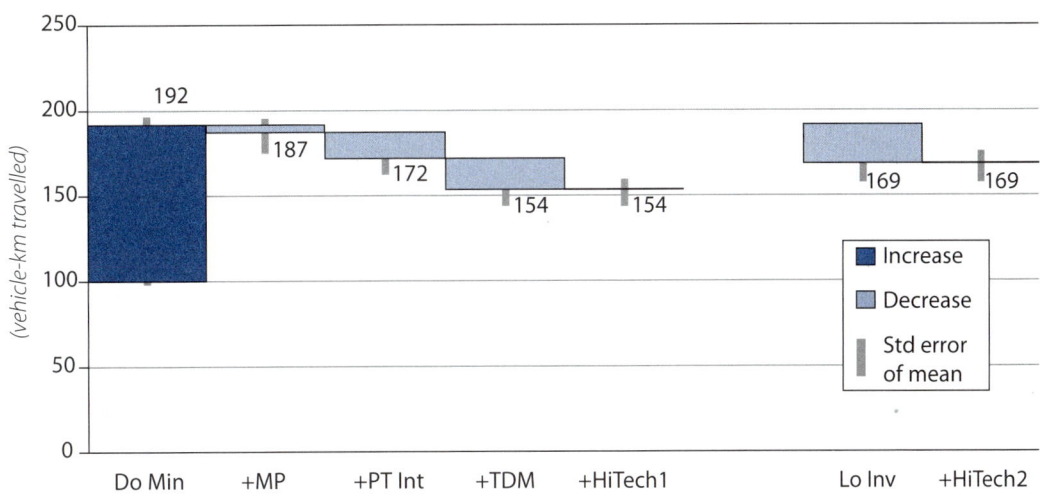

Source: Authors.

Vehicle-Kilometers Travelled

In the "Do Minimal" scenario, the annual vehicle-kilometers travelled (VKT) is forecast to be 192 percent [185 percent–199 percent][53] of that in the base year (figure A1.2).[54] Implementing the master plan makes only a slight difference to the total VKT (187 percent [177 percent–198 percent]). However, better integration of public transport and stronger transport demand managment (TDM) measures make a significant impact. Taken together, they reduce VKT to only 154 percent of the base year figure (integration contributes with [179 percent–172 percent] and TDM with [161 percent–146 percent]). If only a partial (low investment) implementation of the master plan is conducted, a higher VKT results (average 159 percent [159 percent–178 percent]) would be achieved. This analysis is based on an underlying assumption that vehicle operating costs are adjusted such that increasing vehicle fuel efficiency does not impact total VKT, which remains unchanged in the "HiTech" scenarios.

---

53. The numbers in square brackets indicate the likely upper and lower bounds of this average as given by the standard error of the mean. Inside the figures, these are shown as vertical, dotted "whiskers."

54. In figure A1.2 and subsequent figures in the summary, the "Do Minimal" scenario is shown on the figure as "Do Min"; implementation of the complete master plan is shown as "+MP"; additional improved public transport integration is depicted as "+PT Int"; additional transport demand management measures are given as "+TDM"; the higher vehicle technology (fuel efficiency) scenarios are shown as "+HiTech1/2"; and the impact of only a partial (low investment) implementation of the master plan is indicated as "Lo Inv."

Figure A1.3 Fuel Consumption in Base Year *(Kton)*

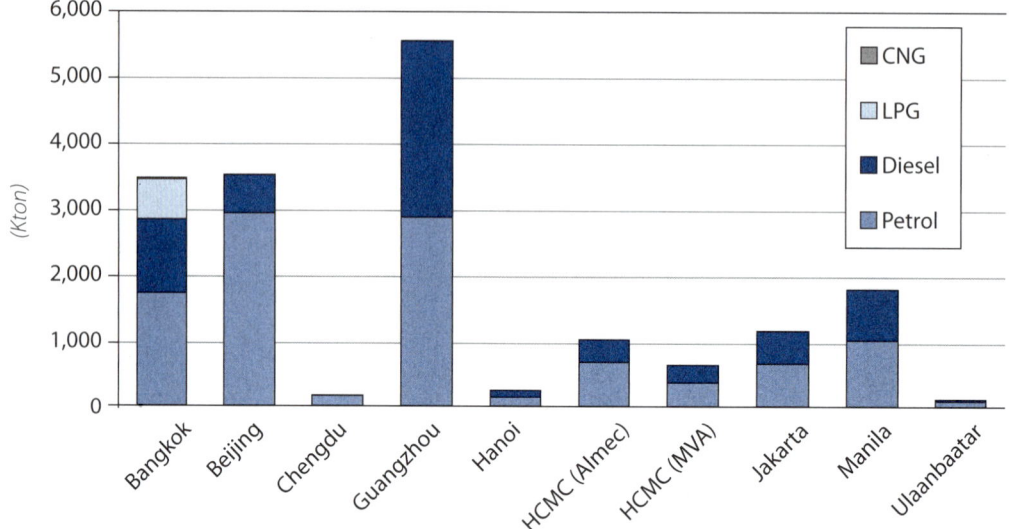

Source: Authors.

Fuel Consumption

The fuel consumption in the base year of each city is given in figure A1.3. The 3 largest cities consumed an average of more than 4.2 million tons of automotive fuel within the study area: Guangzhou (5.6 Mt), Beijing (3.5 Mt), and Bangkok (3.5 Mt). The 3 cities with the lowest annual consumption in the base year were Ulaanbaatar (0.11 Mt), Chengdu (0.16 Mt), and Hanoi (0.24 Mt).

In the "Do Minimal" scenario, the annual fuel consumed by the in-use, on-road vehicle fleet is forecast to be 229 percent [205 percent–253 percent] of that in the base year (figure A1.4). Implementing the master plan (MP) reduces fuel consumption to 214 percent [184 percent–245 percent]. Better integration of public transport and stronger TDM measures together reduce fuel consumption to only 180 percent of the base year figure (integration contributes with [174 percent–216 percent] and TDM with [161 percent–146 percent]). Here the impact of larger buses and trucks can be seen clearly since the fuel consumption increases faster than the total VKT.

An additional "HiTech" scenario is considered that increases light-duty vehicle fuel efficiency in line with current EU plans and proposals. This scenario contemplates new vehicle standards of 130 g/km for passenger cars and 170 g/km for light-duty commercial vehicles in 2015 that are tightened to 100 g/km for passenger cars and 140 g/km for light-duty commercial vehicles in 2020.

This "HiTech" scenario has a clear impact on constraining the increase in total fuel consumption to 148 percent [132 percent–164

Figure A1.4 Total Fuel Consumption Relative to Base Year *(%)*

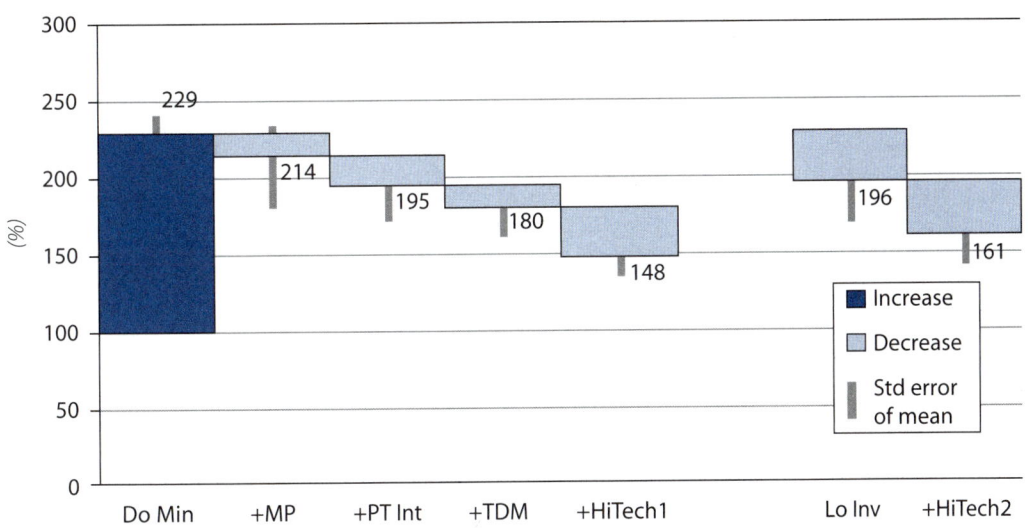

Source: Authors.

percent] of the base year figure. If only a partial (low investment) implementation of the master plan is conducted, the best fuel consumption forecast—including the "HiTech" scenario—is 161 percent [139 percent–184 percent] of the base year's figure.

These growth scenarios are very different for gasoline and for diesel. Under the best case scenario (master plan plus public transport integration, TDM measures, and higher vehicle technology), the consumption of gasoline in the year of the master plan is basically the same as that in the base year [92 percent–114 percent]. However diesel increases by a factor of 250 percent [212 percent–287 percent] (figures A1.5 and A1.6).

In the base year, the total retail value of the automotive fuel consumed in the 10 study areas was US$21.2 billion. This total corresponds to $1.1 billion for 2-wheelers, $10.7 billion for passenger cars and light-duty passenger vehicles, $8.3 billion for goods vehicles, and $1.1 billion for buses.

In the year of the master plan in the "Do Minimal" scenario, total retail automotive fuel consumed increased to $US43.6 billion. Implementing the master plan with better integration of public transport and stronger TDM measures together reduced the cost of fuel consumed to $36.7 billion, saving $6.9 billion. The "HiTech" scenario produced additional fuel savings of $5.6 billion, for a total expenditure on fuel of $31.1 billion. This total corresponds to $0.6 billion for 2-wheelers, $13.8 billion for passenger cars and light-duty passenger vehicles, $14.8 billion for goods vehicles, and $1.9 billion for buses.

Figure A1.5 Gasoline Consumption Relative to Base Year *(%)*

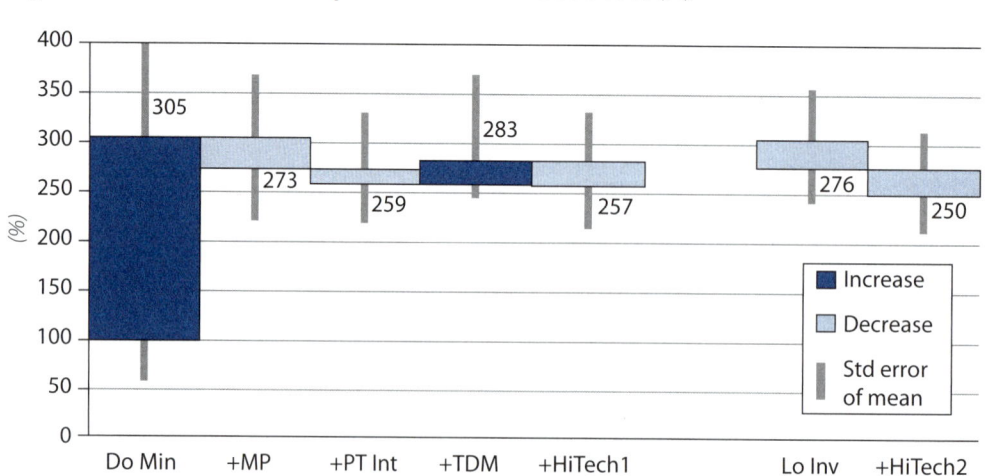

Source: Authors.

CO$_2$ Emissions

CO$_2$ emissions increase in virtually the same proportion as the fuel consumption. The slight difference is due principally to the emerging use of compressed natural gas (CNG) as an urban bus fuel. Under the best case scenario (complete master plan plus public transport integration, TDM measures, and higher vehicle technology), the average CO$_2$ emissions increase to 148 percent [132 percent–165 percent] of those in the base year.

When the new vehicle fuel efficiency standards are not entirely adopted, average CO$_2$ emissions increase to 181 percent [162 percent–200 percent] of those in the base year.

Figure A1.6 Diesel Consumption Relative to Base Year *(%)*

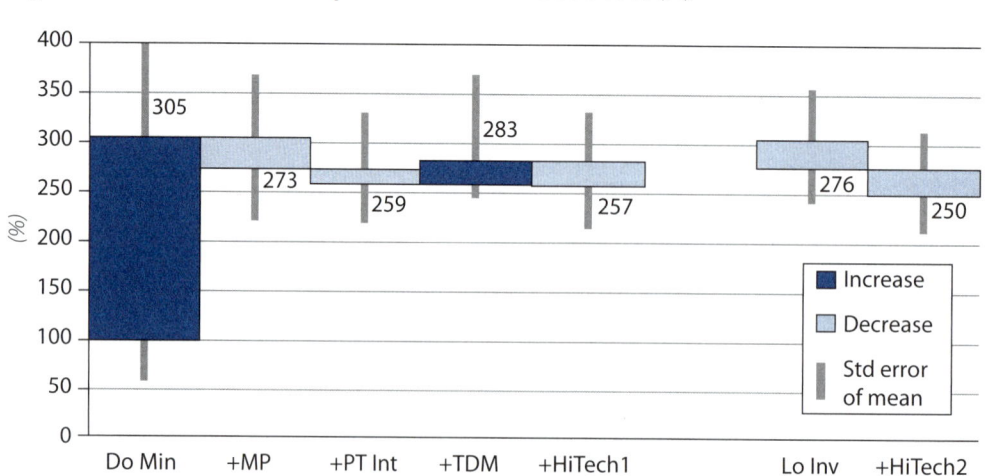

Source: Authors.

Local Pollutants

Figure A1.7 shows the change in average CO emissions and approximate average volatile organic compound (VOC) emissions (on a per-city basis, excepting Ulaanbaatar) for each scenario relative to the base year.

Changing vehicle emissions specifications between the base year and the year of the master plan (up to Euro V) more than compensate the increase in traffic (movement of people and goods) within the urban environment ("Do Min"). The change in specifications reduces CO and VOC emissions by 75 percent compared to the 2007 base values. However, a similar deep reduction does not take place with NOx (figure A1.8), which, in the "Do Min" scenario, remain at 96 percent of their 2007 baseline.

PM2.5 (less than 2.5 microns of particulate matter) emissions fare even worse, showing a "Do Min" scenario value of 12 percent above their 2007 baseline (figure A1.9). However, with the implementation of the master plan, better integration of public transport, and stronger TDM measures combined, as with NOx, PM2.5 emissions end up at approximately 85 percent–87 percent of the 2007 baseline.

In both cases, (NOx and PM2.5), this is due entirely to changing use of heavy duty vehicles and the adoption of only Euro V heavy-duty vehicle emissions standards. Euro VI is focused on the reduction of NOx and PM2.5 from heavy-duty vehicles. Euro VI would limit the value for total oxides of nitrogen (NOx) to 0.4 g/kWh (80 percent less compared with Euro V), and particle mass to 0.01 g/kWh 10 mg/kWh: a 66 percent reduction compared with the Euro V stage limits.

Figure A1.7 CO and VOC Emissions Relative to Base Year *(%)*

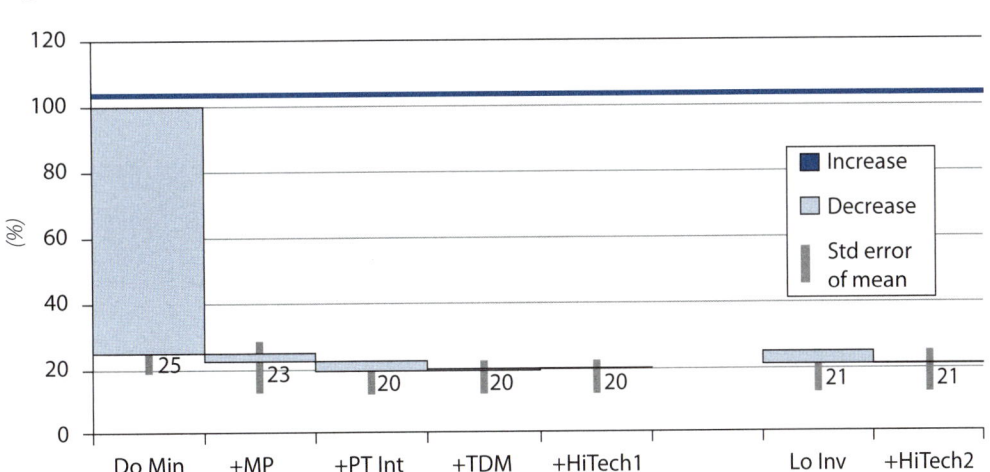

Source: Authors.

Figure A1.8 NO$_x$ Emissions Relative to Base Year *(%)*

Source: Authors.

Figure A1.9 PM2.5 Emissions Relative to Base Year *(%)*

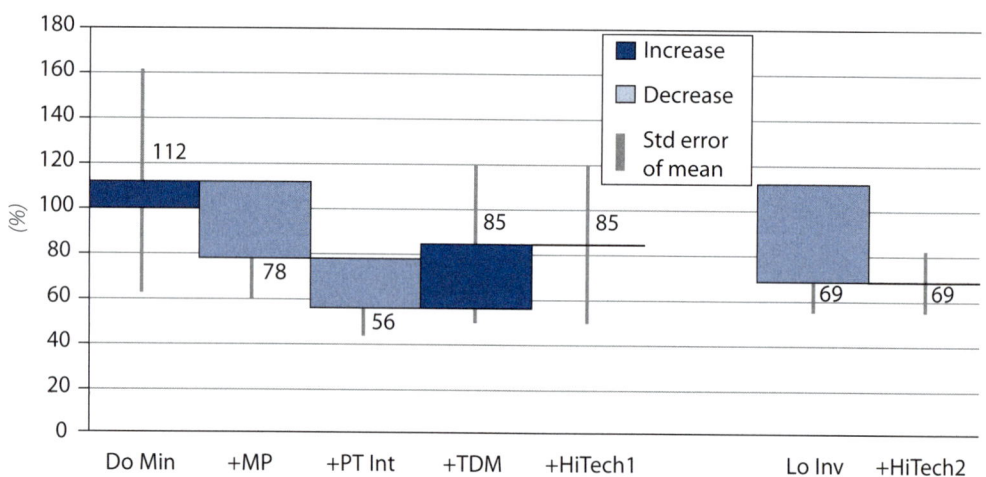

Source: Authors.

Household Study

East Asia is one of the fastest growing regions in the world. From 1990 to 2006, real per capita incomes for the region more than tripled—from $1400 to $4700. This growth is expected to increase in the next few decades. The region also is urbanizing rapidly. By 2030, nearly 53 percent of the population are expected to be living in urban areas. Rising incomes and the transition to urban centers will drive up the demand for appliances in households and increase the demand for electricity consumption.

This parallel household study aims to forecast energy consumption in the region to 2030 and to identify potential energy savings by introducing more energy-efficient appliances. As part of this study, residential energy consumption patterns were evaluated for three countries in the region: Philippines, Thailand, and Vietnam. In each, different levels of action and investment were considered. At one end of the spectrum was the *Baseline (or business-as-usual) scenario*, which predicted how residential electricity consumption might evolve up to 2030 based on market influences if a government did nothing other than what it already had committed to do via existing policies. At the opposite end was the *Alternative (or energy-efficient end-use) scenario*, which modeled a portfolio of technologies through the deployment of new policies that the government could consider in the future.

Key Findings

Economic growth and increased urbanization will drive up household consumption of electricity. The average urbanization rate for the 3 countries studied will be approximately 52 percent by 2030. The Philippines is expected to be the most urbanized with nearly 68 percent of its population living in cities by 2030 (figure A2.1). The mean monthly household expenditures are expected to more than double for urban and rural households.

Under the Baseline scenario, residential energy demand will grow at an average of 4.3 percent per annum to 2030. The Baseline or business-as-usual (BAU) scenario predicts how residential energy consumption could evolve up to 2030 based on current market influences if the governments continue with the existing policies. BAU estimates that the residential sector's annual average electricity consumption will grow at 4.1 percent for Thailand, 4.6 percent for the Philippines, and 4.2 percent for

131

Figure A2.1 Population and Number of Households in the Philippines, Thailand, and Vietnam, 2006–30 *(mil)*

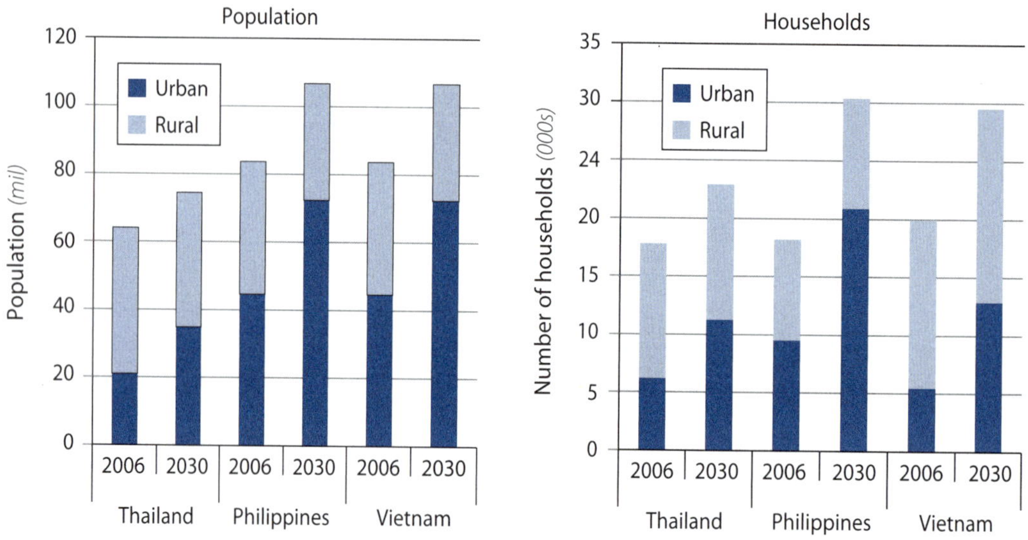

Source: Authors.

Vietnam to 2030. Appliance penetration will be particularly important in driving up demand.

From 2006 to 2030, total electricity consumed by the household sector is expected to more than double. Figure A2.2 shows the total electricity consumption for the 3 countries in 2006 and 2030 under the Baseline scenario. Consumption is expected to grow at an average of 175 percent for the 3 countries over these 24 years.

The main contributor to growth will be increased ownership of high-electricity-consuming devices as household incomes rise. Most of the anticipated growth is expected to be driven by high-energy-consuming appliances for heating/cooling (air conditioners, fans), entertainment (televisions), and kitchen appliances (refrigerators). The average growth rates for these 3 categories over 2006–30 are expected to be 285 percent, 209 percent, and 143 percent, respectively.

The Alternate scenario will engender an average annual savings of 8 percent in energy consumption by 2030. Introducing efficiency improvements in the appliance sector will reduce energy use and gradually will reduce energy consumption. The consumption savings in 2030 are forecast to be the highest for the Philippines (9.5 percent) and somewhat less for the others (7.8 percent for Thailand

Figure A2.2 Household Electricity Consumption in the Baseline Scenario, 2006–30 *(GWh/yr)*

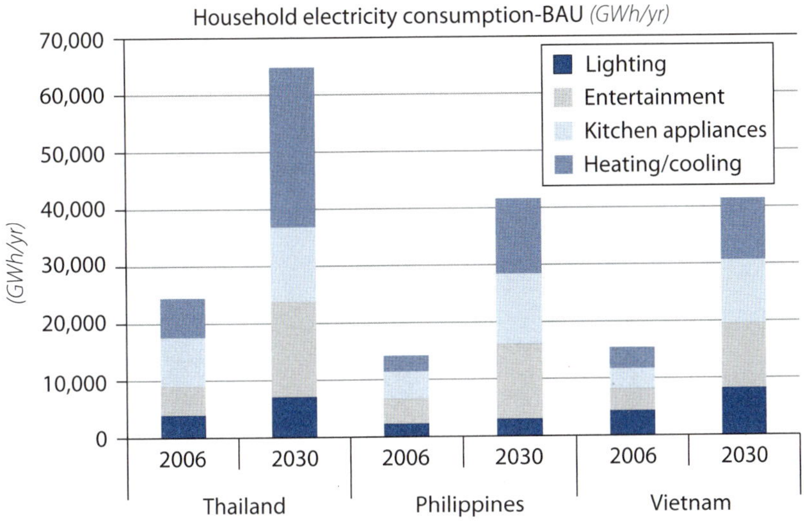

Source: World Bank calculations.

and 7.23 percent for Vietnam). Figure A2.3 depicts the total household energy consumption in 2030 under the 2 scenarios.

Focusing on appliances that account for the bulk of electricity consumption will have a higher impact on energy savings. Potential savings by introducing improvements in efficiency vary across appliance groups (figure A2.4). However, the savings are highest for the high-energy-consuming appliances such as air conditioners, refrigerators, and televisions.

Methodology

Household electric appliance ownership is assumed to depend only on household income. This assumption simplifies the projection of appliance ownership. The rate of electrification was found to be very high for the three countries considered by this study. Hence, only electrified households were used to project appliance ownership and usage.

There are three main steps in forecasting household electricity consumption (figure A2.5). The first step is to project numbers of households and their expenditures in urban and rural areas. The second is to forecast ownership of appliances among the households. The third step computes the electricity consumption, or usage.

Figure A2.3 Household Electricity Consumption in 2030: Baseline vs. Alternate Scenarios *(GWh/yr)*

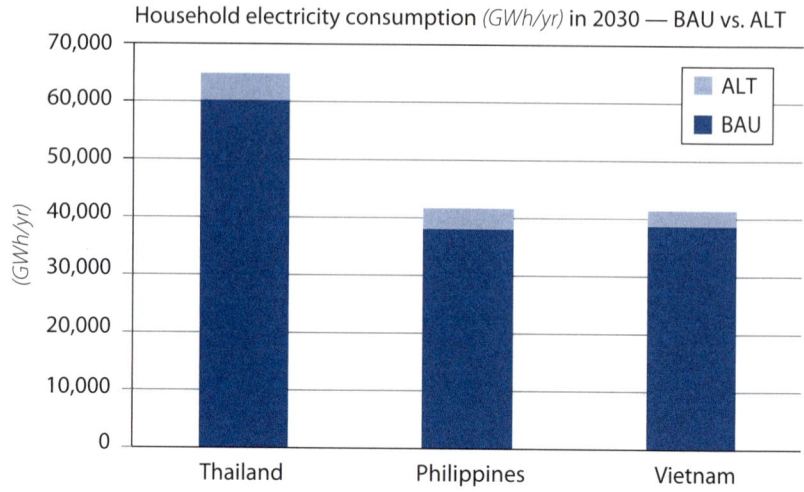

Source: Authors.

Figure A2.4 Energy Savings in 2030 *(%)*

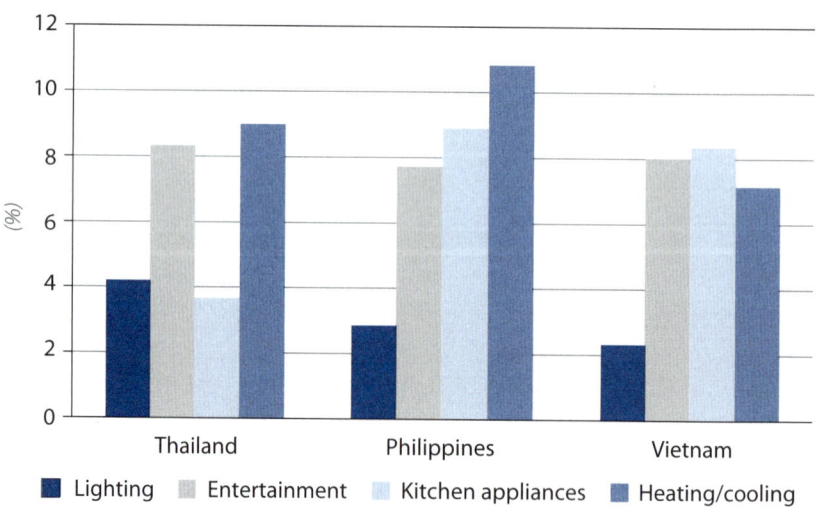

Source: Authors.

Figure A2.5 Modeling Framework: Household Electricity Consumption Module

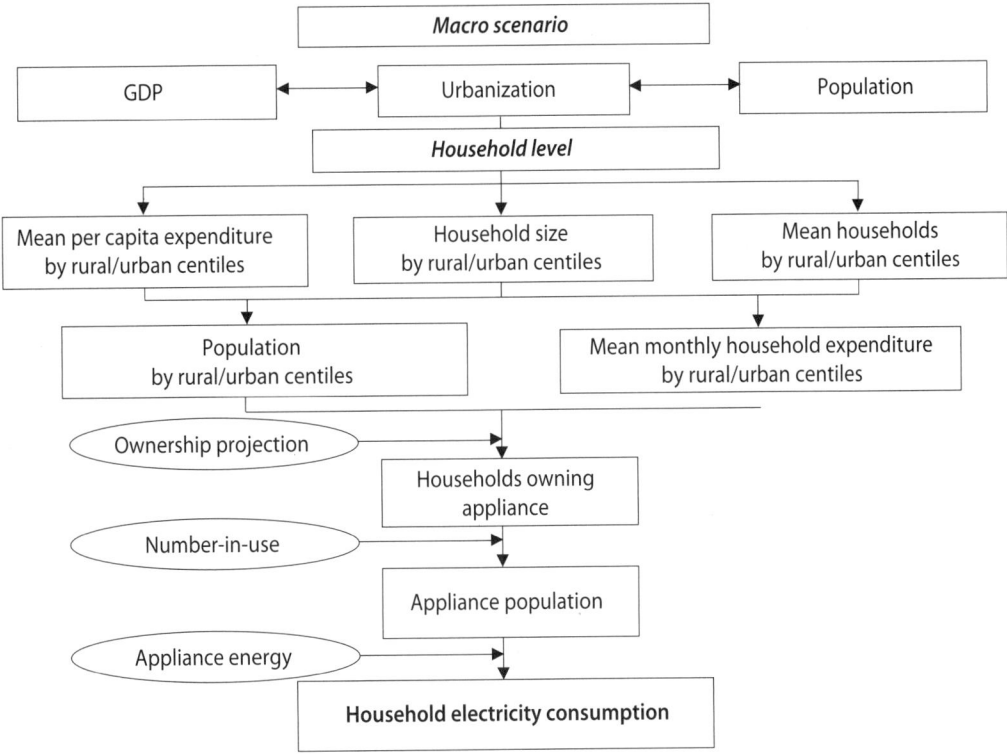

Country Data

Table A3.1 Country Data

Energy	World	EAP	MIC	Cambodia	China	Indonesia
GDP per unit of energy use (2005 PPP $/kg oil equivalent)	5.2	3.4	4.2	4.5	3.2	4.2
Energy use per capita (kg oil equivalent)	1,820.0	1,258.0	1,267.0	351.0	1,433.0	803.0
Energy from biomass products and waste (% of total)	9.8	14.7	12.3	71.3	12.0	29.2
Electric power consumption per capita (kWh)	2,751.0	1,669.0	1,651.0	88.0	2,041.0	530.0
Electricity generated using fossil fuel (% of total)	66.4	82.0	72.9	95.7	82.6	87.8
Electricity generated using hydropower (% of total)	15.9	15.0	20.6	4.1	15.2	7.2
Emissions and pollution						
CO_2 emissions per unit of GDP M(kg/2005 PPP $)	0.5	0.9	0.7	0.0	1.0	0.6
CO_2 emissions per capita (metric tons)	4.5	3.6	3.3	0.0	4.3	1.9
CO_2 emissions growth (%, 1990–2005)	29.5	123.4	43.1	19.5	131.2	181.0
Particulate matter (urban-pop-weighted avg., $\mu g/cu.m$)	50.0	69.0	56.0	46.0	73.0	83.0
Transport sector fuel consumption per capita (liters)	291.0	106.0	144.0	31.0	93.0	118.0

Source: World Bank 2009c.

Japan	Korea, Rep. of	Malaysia	Mongolia	Myanmar	Philippines	Singapore	Thailand	Vietnam
7.5	5.0	4.7	2.6	2.9	6.1	6.5	4.5	3.7
4,129.0	4,483.0	2,617.0	1,080.0	295.0	498.0	6,968.0	1,630.0	621.0
1.3	1.1	4.1	3.8	72.1	26.1	0.0	16.6	46.4
8,220.0	8,063.0	3,388.0	1,298.0	93.0	578.0	8,520.0	2,080.0	598.0
59.2	61.7	92.3	100.0	46.1	64.0	100.0	91.9	58.2
7.9	0.9	7.7	—	53.9	17.5	—	5.9	41.8
0.3	0.4	0.8	1.3	0.3	0.3	0.3	0.6	0.6
9.6	9.4	9.3	3.4	0.2	0.9	13.2	4.3	1.2
13.8	87.2	333.9	(12.0)	165.3	70.7	34.2	182.9	376.0
30.0	35.0	23.0	110.0	58.0	23.0	41.0	71.0	55.0
658.0	534.0	567.0	166.0	28.0	89.0	580.0	314.0	84.0

References

ADB (Asian Development Bank). 2009. "The Economics of Climate Change in Southeast Asia: A Regional Review." Manila. April.

APERC (Asia Pacific Energy Research Centre). 2006. *APERC Energy Demand and Supply Outlook 2006*. APERC under the Institute of Energy Economics.

Berrah, N., F. Feng, R. Priddle, and L. Wang. 2007. *Sustainable Energy in China: The Closing Window of Opportunity*. World Bank.

Blanford, G., R. Richels, and T. Rutherford. 2009. "Feasible Climate Targets: The Roles of Economic Growth, Coalition, Development and Expectations." *Energy Economics* 31 (Supp 2) (December).

BP (British Petroleum). 2009. "BP Statistical Review of World Energy." June. www.bp.com/.../bp.../globalbp/globalbp...energy_review.../2009.../statistical_review_of_world_energy_full_report_2009.pdf

Brown, M.A., F. Southworth, and T.K. Stovall. 2005. *Toward a Climate Friendly Built Environment*. Arlington, VA: Pew Center on Global Climate Change.

Burton, R., D. Goldston, G. Crabtree, L. Glicksman, D. Goldstein, D. Greene, D. Kammen, M. Levine, M. Lubell, M. Savitz, D. Sperling, F. Schlachter, J. Scofield, and J. Dawson. 2008. "How America Can Look within to Achieve Energy Security and Reduce Global Warming." *Reviews of Modern Physics* 80 (4): S1–S109.

Calvin, K., J. Edmonds, B. Bond-Lamberty, L. Clarke, P. Kyle, S. Smith, A. Thomson, and M. Wise. 2009. "Limiting Climate Change to 450 ppm CO_2 Equivalent in the 21st Century." *Energy Economics* 31 (Supp 2) (December).

Capoor, K., and P. Ambrosi. 2009. *State and Trends of the Carbon Market*. World Bank.

China Energy Research Institute. 2009. "2050 China Energy and CO_2 Emission Report." Beijing.

Clarke, L., J. Edmonds, V. Krey, R. Richels, S. Rose, and M. Tavoni. 2009. "International Climate Policy Architectures: Overview of the EMF 22 International Scenarios." *Energy Economics* 31 (Supp 2) (December).

CLASP (Collaborative Labeling and Appliance Standards Program). 2009. Washington, DC. http://www.clasponline.org/index.php

Dahowski, R.T., X. Li, C.L. Davidson, N. Wei, J.J. Dooley, and R.H. Gentile. 2008. "A Preliminary Cost Curve Assessment of Carbon Dioxide Capture and Storage Potential in China." *Energy Procedia* 1 (Issue 1) (February).

Diringer, E., J. Lewis, and L. Clarke. 2008. *Modeling Post-2012 Climate Policy Scenarios*. Washington, DC: Pew Center on Global Climate Change. http://www.pewclimate.org/post2012modeling

EESI (Environmental and Energy Study Institute). 2008. *Jobs from Renewable Energy and Energy Efficiency*. Washington, DC.

Energy Foundation. 2008. *China Energy and Climate Change: Myths, Realities, and Challenges*. San Francisco, CA.

———. 2009. *China Sustainable Energy Program*. San Francisco, CA.

ESMAP (Energy Sector Management Assistance Program). 2006. "Proceedings of the International Grid-Connected Renewable Energy Policy Forum." World Bank.

———. 2008. "An Analytical Compendium of Institutional Frameworks for Energy Efficiency Implementation." World Bank.

———. 2009. "Public Procurement of Energy Efficiency Services." World Bank.

ETAAC (Economic and Technology Advancement Advisory Committee). 2008. *Technologies and Policies to Consider for Reducing Greenhouse Gas Emissions in California*. Sacramento, CA.

Gibbins J., and H. Chalmers. 2008. "Preparing for Global Rollout: A 'Developed-Country-First' Demonstration Program for Rapid CCS Deployment." *Energy Policy* 36 (2008): 501–07.

Goldstein, D.B. 2007. *Saving Energy, Growing Jobs: How Environmental Protection Promotes Economic Growth, Profitability, Innovation, and Competition*. Berkeley, CA: Bay Tree Publishing.

Hosier R., N. Kulichenko, A. Maheshwari, N. Toba and X. Wang, 2010. "Financing Instruments for Climate Change Mitigation: Strategic Use for Expanded Impacts."

IEA (International Energy Agency). 2006. *World Energy Outlook 2006*. Paris.

———. 2007. *World Energy Outlook 2007*. Paris.

———. 2008a. *World Energy Outlook 2008*. Paris.

_____. 2008b. *Energy Technology Perspective 2008: Scenarios and Strategies to 2050*. Paris.

_____. 2008c. "Energy Statistics of Non-OECD Countries." Paris.

_____. 2008d. "Energy Balances of Non-OECD Countries." Paris.

_____. 2008e. *Worldwide Trends in Energy Use and Efficiency*. Paris.

_____. 2008f. *CO_2 Capture and Storage: A Key Carbon Abatement Option*. Paris.

_____. 2008g. *Empowering Variable Renewables: Options for Flexible Electricity Systems*. Paris.

IIASA (International Institute for Applied System Analysis). 2009. "GGI Scenario Database." Laxenburg, Austria.

IMF (International Monetary Fund). 2006. IMF World Energy Outlook. Washington, DC.

IPCC (Intergovernmental Panel on Climate Change). 2007. *Climate Change 2007: Mitigation. Contribution of Working Group III to the Fourth Assessment Report of the Intergovernmental Panel on Climate Change*. Cambridge and New York.

Kats, G. 2008. *Greening Buildings and Communities: Costs and Benefits*. London: Good Energies.

Keystone Center. 2007. *Nuclear Power Joint Fact-Finding*. Keystone, CO.

Kenworthy, J., and F. Laube. 2002. "Urban Transport Patterns in a Global Sample of Cities and Their Linkages to Transport Infrastructures, Land Use, Economics and Environment." *World Transport Policy and Practice* 8 (3): 5–20.

Knopf, B., O. Edenhofer, T. Barker, N. Bauer, L. Baumstark, B. Chateau, P. Criqui, A. Held, M. Isaac, M. Jakob, E. Jochem, A. Kitous, S. Kypreos, M. Leimbach, B. Magne, S. Mima, W. Schade, S. Scrieciu, H. Turton, and D. van Vuuren. 2010. "The Economics of Low Stabilisation: Implications for Technological Change and Policy." In M. Hulme and H. Neufeldt (eds.), *Making Climate Change Work for Us*. Cambridge University Press.

Lin J., N. Zhou, M. Levine, and D. Fridley. 2006. *Achieving China's Target for Energy Intensity Reduction in 2010: An Exploration of Recent Trends and Possible Future Scenarios*. Lawrence Berkeley National Laboratories, University of California Berkeley, CA.

Martinot, E., and J. Li. 2007. *Powering China's Development*. Washington, DC: Worldwatch.

Mao, J. 2009. "How China Reduces CO_2 Emissions from Coal-Fired Power Generation." Prepared for Energy Week. World Bank.

McKinsey Global Institute. 2008. *China's Green Revolution*. Washington, DC.

———. 2009a. *Promoting Energy Efficiency in the Developing World*. Washington, DC.

———. 2009b. *Toward a Global Climate Agreement: Project Catalyst*. Washington, DC: ClimateWorks Foundation.

Meyers, S., J. McMahon, and M. McNeil. 2005. *Realized and Prospective Impacts of U.S. Energy Efficiency Standards for Residential Appliances: 2004 Update*. Berkeley, CA: Lawrence Berkeley National Laboratory, University of California.

MIT (Massachusetts Institute of Technology). 2003. *The Future of Nuclear Power: An Interdisciplinary MIT Study*. Cambridge, MA.

National Bureau of Statistics of the People's Republic of China. 2006. *China Statistical Yearbook 2006*. Beijing: China Statistics Press.

NRDC (National Resources Defense Council). 2007. *The Next Generation of Hybrid Cars: Plug-in Hybrids Can Help Reduce Global Warming and Slash Oil Dependency*. Washington, DC.

Philibert, C. 2007. *Technology Penetration and Capital Stock Turnover: Lessons from IEA Scenario Analysis*. Paris: International Energy Agency (IEA). May.

Rao, S., K. Riahi, E. Stehfest, D. van Vuuren, C. Cho, M. den Elzen, M. Isaac, and J. van Vliet. 2008. *IMAGE and MESSAGE Scenarios Limiting GHG Concentration to Low Levels*. Laxenburg, Austria: International Institute for Applied Systems Analysis.

REN21. 2008. *Renewables 2007 Global Status Report*. Paris and Washington: Renewable Energy Policy Network for the 21st Century Secretariat and Worldwatch Institute.

Riahi, K., A. Grubler, and N. Nakicenovic. 2007. "Scenarios of Long-Term Socio-Economic and Environmental Development under Climate Stabilization." *Technological Forecasting and Social Change* 74 (7): 887–935.

Rubin, E. 2008. "Financing Carbon Capture and Sequestration." Presentation at the World Bank Clean Coal Initiative, June 2008.

Sarkar, A., and J. Singh. 2009. "Actions to Scale up Energy Efficiency Implementation in Developing Countries: Compendium of Ideas."

Shalizi, Z., and F. Lecocq. 2008. "Economics of Targeted Mitigation Programs in Sectors with Long-lived Capital Stock." Background paper for *World Development Report 2010*. World Bank.

Sorrell, S. 2008. "The Rebound Effect: Mechanisms, Evidence and Policy Implications." Paper presented at the Electricity Policy Workshop, Toronto.

Stern, N. 2006. *Stern Review on the Economics of Climate Change*. Cambridge University Press.

Sterner, T. 2007. "Fuel Taxes: An Important Instrument for Climate Policy." *Energy Policy* 35: 3194–202.

Taylor, R.P., C. Govindarajalu, J. Levin, A.S. Meyer, and W.A. Ward. 2008. *Financing Energy Efficiency: Lessons from Brazil, China, India and Beyond*. World Bank.

United Nations. 2007a. *State of the World Population 2007: Unleashing the Potential of Urban Growth*. New York: United Nations Population Fund (UNFPA).

———. 2007b. "World Population Prospects: World Urbanization Prospects. The 2007 Revision." ECOSOC (Department of Economic and Social Affairs of the United Nations Secretariat), Population Division. New York. http://esa.un.org/unup

WBCSD (World Business Council for Sustainable Development). 2008. *Power to Change: A Business Contribution to a Low-Carbon Economy*. Geneva.

Weber, C.L., G.P. Peters, D. Guan, and K. Hubacek. 2008. "The Contribution of Chinese Exports to Climate Change" *Energy Policy* 26: 3572–77. London: Elsevier.

WEF (World Economic Forum). 2009. *Green Investing: Toward a Clean Energy Infrastructure*. Geneva.

World Bank. 2006a. Indonesia Poverty Asssment

———. 2006b. *Renewable Energy Toolkit: A Resource for Renewable Energy Development*.

———. 2008a. *The Development of China's ESCO Industry, 2004–2007*.

———. 2008b. *Mid-term Evaluation of China's 11th Five-Year Plan*.

———. 2008c. *An Evaluation of World Bank Win-Win Energy Policy Reforms*.

———. 2008d. *Energy Efficiency in Middle East and North Africa*.

———. 2008e. *Accelerating the Development and Commercialization of Advanced Energy Technologies in Developing Countries*.

————. 2009a. *World Development Report 2010: Development and Climate Change.*

————. 2009b. *Beyond Bonn: World Bank Group Progress on Renewable Energy and Energy Efficiency in Fiscal 2005–2009.*

————. 2009c. *World Development Indicators.*

————. 2009d. *Battling the Forces of Global Recession.*

————. 2009e. *Indonesia Low Carbon Study.*

————. 2009f. *Expanding Opportunities for Improving Energy Efficiency in Vietnam.*

————. 2009g. *Coal Power Technology Country Study China and India.*

————. 2009h. "Eco2 Cities: Ecological Cities as Economic Cities."

Worldwatch Institute. 2009. *State of the World 2009: Into a Warming World.* New York and London: W.W. Norton.

WRI (World Resource Institute). 2005. Climate Analysis Indicators Tool (CAIT) vers. 3.0. Washington, DC. http://cait.wri.org

Zhang, X. 2008. *Observations on Energy Technology Research, Development and Deployment in China.* Beijing: Tsinghua University Institute of Energy, Environment and Economy.

Index

Boxes, figures, notes, and tables are indicated by b, f, n, and t following the page numbers.